D1462181

FALLING BEHIND?

FALLING BEHIND?

Boom, Bust, and the Global Race for Scientific Talent

Michael S. Teitelbaum

Princeton University Press
Princeton and Oxford

Published by Princeton University Press,
41 William Street, Princeton, New Jersey 08540

In the United Kingdom: Princeton University Press,
6 Oxford Street, Woodstock, Oxfordshire OX20 1TW
press.princeton.edu

Jacket Photograph: Details of photographs from the Momentum Series
© Alejandro Guijarro

Jacket design by Jessica Massabrook

ISBN 978-0-691-15466-4
Library of Congress Control Number: 2013957467

British Library Cataloging-in-Publication Data is available

This book has been composed in Sabon Next LT Pro
and Akzidenz-Grotesk BQ

Printed on acid-free paper ∞
Printed in the United States of America

10 9 8 7 6 5 4 3 2 1

CONTENTS

ACKNOWLEDGMENTS

I am very grateful for the assistance provided by organizations and colleagues in the development of this book. In particular the Labor and Worklife Program at Harvard Law School and the Alfred P. Sloan Foundation provided important support at critical times. Many individuals too numerous to mention have also provided generous assistance, suggestions, and constructive criticism. Special thanks are due to Richard Freeman, Paula Stephan, Elaine Bernard, Jack Trumphour, Alida Castillo Freeman, Lorette Baptiste, Walter Schaffer, Nirmala Kannankutty, Mark Regets, James Lightbourne, Philip Martin, Sharon Stanton Russell, David Kaiser, Bruce Morrison, Hal Salzman, Lindsay Lowell, Vivien Stewart, Paul Joskow, Ralph Gomory, Daniel Goroff, Clinton Oxenrider, Robin Wagner, Michael Boyle, Robert Wyman, Ronil Hira, and Henry Bourne. The project also benefited from the valuable advice of Seth Ditchik, Beth Clevenger, and Brigitte Pelner at Princeton University Press, and copyeditor Karen Verde.

Of course the book's analyses should not be interpreted as necessarily those of the institutions and individuals mentioned above. These, and any factual errors, are entirely my responsibility.

FALLING BEHIND?

Introduction

In this increasingly globalized world, respected and influential voices warn urgently that the United States is *falling behind* in a global "race for talent" that will determine the country's future prosperity, power, and security. Expressions of such concerns have become common, even conventional, and are embraced with little question by many who have leadership roles in politics, business, media, and education. The gist of this perspective and its key assumptions might be fairly summarized as follows:

> The "second wave" of globalization now under way differs significantly from the "first wave" of about a century ago.[1] Now a nation's economic prosperity is no longer closely related to its physical capital, natural resources, and economic system, but instead is driven by its "human capital." It is the education, skill, creativity, and entrepreneurship of a country's population that will determine whether it will prosper or fall behind in the twenty-first century.
>
> The dominant economic role now being played by science and technology means that the core of any nation's human capital consists of the size and creativity of its science and engineering workforce. Hence it is critical for the future of the United States (and indeed of all nations) both to educate domestically *and* to attract from abroad the largest feasible numbers of the "best and brightest" of scientists and engineers. These resources of critical human capital will, in turn, propel the economic growth and prosperity of the nation. Countries that fall behind in science and technology will stagnate economically as others charge forward. Moreover, leading-edge capabilities in science and engineering also have become central to every nation's international and domestic security.
>
> In short, scientists and engineers form the vanguard of each country's future "competitiveness" and security in a globalized world.

For some, the subject is more than an issue of competitive advantage among nations in economic or security terms. Indeed, it is a matter of global human survival, expressed in terms approximating the following:

Humanity as a whole has much to gain from collective investments in human capital in science and engineering. Research in basic biomedical science is the wellspring of major advances against diseases such as cancer, HIV, malaria, and new epidemics. The creativity of scientists and engineers in biomedical fields enables reduced mortality and healthier lives for all of humanity, lower expenditures on healthcare, and more productive workforces worldwide. Scientists and engineers in other fields are of equal importance to the future of humanity, advancing understanding and capabilities in chemistry, physics, energy, and earth sciences that contribute to the global good by enhancing collective understanding of Earth's environment and of effective means for mitigating damage to it.

Guided by such perspectives, many corporate, political, and opinion leaders in the United States have been sounding persistent alarms about current or future "shortages" in the nation's human capital in science and engineering, and more generally to unfavorable trends relative to those in other countries. If their concerns can be encapsulated in a single sentence, it might read as follows:

The United States, long a leader in the number and quality of its scientists and engineers, has been *falling behind* its international competitors, and is thereby risking serious deterioration in its future prosperity and security.

These recent alarming assessments of the state of U.S. education and research in science and engineering turn out to be quite inconsistent with a very substantial body of research literature produced by independent scholars. Nonetheless, the U.S. political system during the past decade clearly has been highly responsive to claims of "shortages" or "shortfalls" of scientists and engineers, and has taken actions designed to increase the number of scientists and engineers in the U.S. workforce.

This political responsiveness to such assertions of alarm is by no means a new phenomenon. Quite the contrary: concern about "shortages" has a long and fascinating history that goes back at least to World War II. It is a story that lies at the heart of many of the central domestic and international developments, both political and economic, of that tumultuous period. Perversely, past shortage claims, some of which are eerily similar to those being heard today, have led to repeated three-stage cycles of *alarm, boom*, and *bust* that have buffeted and destabilized the nation's science and engineering workforce.

In stage 1 of such cycles, the *alarm* has been sounded about the United States "falling behind" in the supply of scientists and/or engineers. In stage 2, the U.S. political system has responded to these alarms with measures that generated rapid expansion in the supply of scientists and engineers. This stage 2 *boom* has then generally (though not always) been followed by stage 3 of the cycle—a *bust* in which expanded numbers of enthusiastic young scientists and engineers, some of whom had devoted many years to advanced education, unexpectedly have found themselves facing chilly labor markets and unattractive career prospects. Finally the cycle has come full circle, as knowledge of the unhappy career experiences of recent graduates cascaded down to talented younger generations of U.S. students who have chosen to pursue other career paths, thereby stimulating a new round of alarms about impending shortages.

This is hardly a happy or uplifting history. But it is a history from which much could be learned to inform competing claims that are readily apparent in current controversies about the prospects for U.S. science and engineering.

In Brief, the Three Core Findings of This Book

The evidence assembled in this book leads inescapably to three core findings:

- First, that the alarms about widespread shortages or shortfalls in the number of U.S. scientists and engineers are quite inconsistent with nearly all available evidence;

- Second, that similar claims of the past were politically successful but resulted in a series of booms and busts that did harm to the U.S. science and engineering enterprise and made careers in these fields increasingly unattractive; and

- Third, that the clear signs of malaise in the U.S. science and engineering workforce are structural in origin and cannot be cured simply by providing additional funding. To the contrary, recent efforts of this kind have proved to be destabilizing, and advocates should be careful what they wish for.

The book is organized as follows. In chapter 1, we review several recent politically influential reports, all of which emphasized the critical need for legislation and public expenditures to increase the number of scientists and engineers entering the U.S. workforce. The discussion assesses the data and analyses underlying these reports and the overlap among the constituencies that produced them.

In chapter 2, we discuss a half-century of experience with earlier influential reports that urged similar actions in prior decades. The chapter discusses no fewer than five earlier rounds of such concerns that go back to the late 1940s. Each cycle lasted for between 10 and 20 years, and generally followed the same three-stage pattern of "alarm/boom/bust."

> *Round 1* began in the decade immediately following World War II. The focus by the U.S. government in this period was on large increases in the number of physicists, seen as a strategic human resource essential to Cold War competition with the Soviet Union. By the mid-1950s, the number of recent PhDs in physics had grown very rapidly, but unexpectedly those newly emerging graduates were beginning to experience difficult career prospects. In this case a full-blown bust seems likely to have ensued had it not been for the launching of Sputnik 1 in October 1957 that initiated Round 2.

> *Round 2*, driven by political shock over the Sputnik launches, produced even larger increases in the U.S. science and engineering workforce. By the late 1960s, however, political enthusiasm had waned sharply for federal funding of science and engineering, producing an ensuing bust of serious magnitude in the 1970s.

> *Round 3* was driven by several federal initiatives—the "war on cancer" that had begun in the 1970s, the 1980s defense buildup under President Reagan, and anxious reports from federal agencies during the 1980s. A 1983 federal commission report described "A Nation at Risk" because of a failing public education system, and a few years later other federal reports sounded alerts about "looming shortfalls" of scientists and engineers. Again increased government funding was provided to expand the number of scientists and engineers. By the late 1980s, however, an economic recession and the collapse of the Soviet Union led to declines in spending on science and engineering and reversal of Reagan's defense buildup, all contributing to an ensuing bust in the early 1990s.

> *Rounds 4 and 5*, discussed in part 2 of chapter 2, took place after the end of the Cold War and so lacked the national security elements of the earlier three rounds. Rounds 4 and 5 had different origins, but overlapped in time—Round 4 ran roughly from 1995 to 2005, while Round 5 covered the years 1998–2008.

> The origins of *Round 4* lay in powerful and concurrent booms in several high-tech industries (especially information technology, Internet, telecommunications, and biotech), along with a brief episode of large-scale expenditures to "fix" critical software that many warned might fail due to the impending end of the twentieth century, and hence known as the "Year 2000," "Y2K," or "Millennium

bug" problem.[2] These concurrent booms were followed by concurrent busts in all of these industry sectors beginning around 2001. Round 4 also initiated a new strategy that persists to the present day. Coupled with the waning of national security concerns driven by the Cold War, the new availability of large pools of scientists and engineers in low-income countries such as China and India led U.S. employers to advocate successfully for expanded access to large numbers of foreign workers admitted on temporary visas.

Round 5 affected only biomedical research, driven by a successful lobbying effort warning of inadequate federal funding for such research. In response, the federal government sharply increased biomedical research funding by (literally) doubling the budget of the National Institutes of Health over a five-year period from 1998 to 2003. By the end of this period, though, political enthusiasm for further increases had waned as budget constraints emerged and members of Congress in key positions changed. Subsequent NIH budgets were essentially flat, but even in the absence of large cuts these flat budgets produced a sudden bust variously described as a "hard landing" or a true "funding crisis." This bust was later moderated temporarily by a massive infusion of short-term funds in 2009 and 2010, as part of the unexpected economic stimulus package to counteract the economic emergency that began in 2008, only to return to renewed alarms about insufficient federal funding for biomedical research.

In chapter 3, we explore the question of why these repeated cycles of *alarm/boom/bust* have occurred, and assess whether in the end they have mattered. The producers of the studies and reports related to the earlier cycles—which came to widely differing conclusions—were many and various: government agencies such as the National Science Foundation, Department of Commerce, and the Government Accountability Office; nonprofit analytic organizations such as RAND, National Research Council, and Urban Institute; employer organizations such as the Information Technology Association of America and the Business Roundtable; corporations seeking political support for their views; advocacy groups producing their own advocacy "research"; as well as independent academic researchers in a number of universities. For the most influential of these reports, we offer detailed case studies describing the origins, personnel, funding, and promotional efforts underlying each.

In chapter 4, we consider the influential roles played by interest groups and their lobbyists in these cycles. Which such groups have been most unified or most divided, most influential or unsuccessful in their efforts? To what extent have interest groups effectively used credible empirical evidence and research, or to what extent used "advocacy research" of little credibility other than in the political domain?

In chapter 5, we explore the unique characteristics of labor markets for scientists and engineers. How do these characteristics affect public perceptions, and to what extent have the successive cycles of *alarm/boom/bust* affected the attractiveness of careers in these fields? Public discussion has been dominated by persistent but contradictory claims of "shortages" and "surpluses" of scientists and engineers. These emanated from employers and their organizations, higher education, think tanks and independent experts on U.S. labor markets, government agencies, and the media—in many cases these groups have been talking past one another.

In chapter 6 we describe in some detail the distinctive structures that "produce" most of the country's scientists and engineers, along with increasing fractions of those from other countries. These structures, most of which have evolved since World War II, include the world-class array of U.S. research universities, vast governmental funding agencies such as the National Institutes of Health and the National Science Foundation, and the intersections between these structures with the remarkably complex U.S. legal system that is supposed to regulate migration of permanent immigrants, temporary workers, and international students.

Chapter 7 focuses on the U.S. science and engineering workforce in international comparison, addressing in particular some of the recent trends and patterns that have evoked expressions of both concern and confidence in the United States about "competitiveness."

In chapter 8, we conclude with an overall assessment of the U.S. system that has evolved as the joint driver of both basic research and higher education in science and engineering. To what extent have the outputs of this system been successful? Have they been significant positive forces in the economic development and prosperity of the United States? Have some features of this system evolved in ways that are counterproductive? If so, how might the current structure be incrementally modified or tweaked both to maximize the positive and minimize the negative? We consider whether the repeated alarms sounded over the past six decades may be the only way to gain high-level political attention to the important policy issues surrounding science and engineering. Finally, we also discuss whether changes to this system are feasible, or in the alternative more likely to be effectively blocked by those whose interests would lead them to resist.

CHAPTER 1
Recent Alarms

In the race for the future, America is in danger of falling behind ... our generation's Sputnik moment is back.

—President Barack Obama, 2010, "Remarks by the President on the Economy in Winston-Salem, North Carolina," December 6, 2010

Three highly influential reports, all released within a five-month period in 2005 and all guided by prominent corporate leaders, have dominated the past years of discussions about whether the United States is falling behind in terms of its science and engineering workforce. These three followed different styles but had much in common, and for good reason, as we shall see.

The first report, entitled *Innovate America*, was published in May 2005 as a product of the "National Innovation Initiative" of the Council on Competitiveness; it addressed a very broad range of issues it considered central to innovation. The second and third of the reports published in 2005 focused heavily upon the issues surrounding the Science, Technology, Engineering, and Mathematics (STEM) workforce. *Tapping America's Potential* (TAP)[1] was produced and published in July 2005 by the Business Roundtable, an association of CEOs of large U.S.-headquartered corporations. The last of this report trio, released in October 2005, was produced by an ad hoc committee appointed by the National Research Council and bore the evocative title *Rising Above the Gathering Storm: Energizing and Employing America for a Brighter Economic Future* (Gathering Storm).[2]

Both the TAP and Gathering Storm reports recounted indicators of decline in both the quantity and quality of U.S. students graduating from the nation's K-12 primary and secondary education systems, particularly their skills in science and mathematics. Both made the case that the result is inadequate numbers of scientists and engineers—whether current or projected—that pose profound threats to the future of U.S. economic prosperity and security.

The views of leaders of corporations, business associations, and research universities that energized all three of these 2005 reports were echoed and amplified by prominent journalists and editorial writers; by leaders of K-12 education; by prominent figures in higher education and research; by numerous state governors; and by national politicians of both parties. Indeed, it is fair to say this perspective has been and continues to be the conventional and dominant view among elite U.S. opinion leaders.

Yet, as we shall see, it is also a perspective that has been but little scrutinized in an objective way, and rarely tested against empirical evidence. It is the goal—perhaps the overly ambitious goal—of this book to describe what is known, what is unknown, and even what is intrinsically unknowable about this critical set of issues.

Innovate America

This report, produced by a project called the National Innovation Initiative organized by the Council on Competitiveness, was led by a nineteen-member "Principals Committee." This committee was comprised of ten CEOs of major corporations and nine presidents of leading research universities and institutions, and was co-chaired by Samuel J. Palmisano, CEO of IBM Corporation and G. Wayne Clough, president of Georgia Institute of Technology (see table 1.1).

A related advisory committee was co-chaired by Norman R. Augustine, retired CEO of Lockheed Martin Corporation and William R. Brody, president of Johns Hopkins University. The report lists numerous working groups in addition to these leadership committees, and hundreds attended the "National Innovation Initiative Summit" in December 2004 to discuss the Initiative's recommendations.[3]

The scope of the *Innovate America* report was far broader than the two reports that followed and are discussed later, as it addressed the entire "innovation ecosystem" of the U.S. economy. Its recommendations included improvements in U.S. "infrastructure," including support for innovative manufacturing, national prizes for innovation, improvements to the U.S. patent system, and expansion of integrated health data systems. In addition there were recommendations under the heading "investment" that included expanded federal support for the physical sciences and engineering, a permanent and restructured research and development (R&D) tax credit for corporations, increased tax incentives favoring early-stage risk capital provided

Table 1.1. Members of Principals Committee, National Innovation Initiative

Samuel J. Palmisano, Co-Chair
Chairman and Chief Executive Officer, IBM Corporation
G. Wayne Clough, Co-Chair
President, Georgia Institute of Technology
Gerard J. Arpey
Chairman, Chief Executive Officer and President, AMR and American Airlines
Lee C. Bollinger
President, Columbia University
Molly Corbett Broad
President, University of North Carolina
Michael J. Burns
Chairman, President and Chief Executive Officer, Dana Corporation
Mary Sue Coleman
President, University of Michigan
Denis A. Cortese
President and Chief Executive Officer, Mayo Clinic
The Honorable Robert M. Gates
President, Texas A&M University
Sheryl Handler
Chief Executive Officer, Ab Initio
John L. Hennessy
President, Stanford University
The Honorable Shirley Ann Jackson
President, Rensselaer Polytechnic Institute
Vikram Pandit
President and Chief Operating Officer, Institutional Securities and Investment Banking Group, Morgan Stanley
Steven S. Reinemund
Chairman of the Board and Chief Executive Officer, PepsiCo, Inc.
W. J. Sanders III
Founder and Chairman Emeritus, Advanced Micro Devices, Inc.
Ivan G. Seidenberg
Chairman and Chief Executive Officer, Verizon
Kevin W. Sharer
Chairman, Chief Executive Officer, and President, Amgen, Inc.
Charles M. Vest
President, Massachusetts Institute of Technology
G. Richard Wagoner, Jr.
Chairman and Chief Executive Officer, General Motors Corporation

by angel networks and seed capital funds, and reforms in the U.S. tort system. Under its third main heading of "talent," *Innovate America* noted that K-12 education was not its primary focus,[4] but did make recommendations for U.S. higher education including federal funding for at least 5,000 new portable graduate fellowships in science and engineering, tax deductions for private sector scholarships for U.S. undergraduates in science and engineering, expansion of Professional Science Master's programs at U.S. universities, and measures to attract international science and engineering students and provide them with work permits. As we shall see, these latter recommendations had much in common with those embraced by the two later reports issued by other organizations that same year.

Tapping America's Potential (TAP)

The TAP report, a declarative pamphlet only nineteen pages long, was produced in July 2005 by a coalition of industry associations led by the Business Roundtable, "an association of chief executive officers of leading U.S. companies with more than over $6 trillion in annual revenues and more than 14 million employees."[5] Its signatories included fourteen other politically influential business organizations such as the National Association of Manufacturers and the U.S. Chamber of Commerce.

The report was addressed "To Leaders Who Care about America's Future." It began with an expression of "deep concern" about the ability of the United States to sustain its leadership in science and technology and thereby to maintain its economic competitiveness. In response to such concerns, it called for a rapid doubling of the number of science, technology, engineering, and mathematics graduates earning bachelor's degrees during the decade from 2005 to 2015.

Its perspective and recommendations were succinctly summarized in its first few paragraphs:

> Fifteen of our country's most prominent business organizations have joined together to express our deep concern about the United States' ability to sustain its scientific and technological superiority through this decade and beyond. To maintain our country's competitiveness in the 21st century, we must cultivate the skilled scientists and engineers needed to create tomorrow's innovations.
>
> *Our goal is to double the number of science, technology, engineering and mathematics graduates with bachelor's degrees by 2015.*[6]
>
> The United States is in a fierce contest with other nations to remain the world's scientific leader. But other countries are demonstrating a greater commitment to building their brainpower.

The TAP report began with the ominous (and factually correct) observation that "History is replete with examples of world economies that once were dominant but declined because of myopic, self-determined choices." It then focused on what it called "the critical situation in U.S. science, technology, engineering and mathematics." It pointed to numerous "warning signs," including waning achievement and interest in science and mathematics among U.S. students; higher interest in science and engineering among competitor nations such as China; rising production of engineers in such countries; increasing dependence in the United States on foreign-born scientists and engineers; and lagging government support for basic research in the physical sciences.

The report argued that it is essential for the United States to maintain its competitiveness in the twenty-first century, and that to do so it must create a new "National Education for Innovation Initiative," a "21st-century version of the post-Sputnik national commitment to strengthen science, technology, engineering and math education." To this end it urged a "public/ private partnership to promote, fund and execute a new National Education for Innovation Initiative ... [that] must be broader than the 1958 [post-Sputnik] National Defense Education Act because federal legislation is only one component of a larger, more comprehensive agenda."[7] A primary goal would be to enhance the attractiveness of K-12 science and math teaching as a career, so as to "cultivate the skilled scientists and engineers needed to create tomorrow's innovations."[8]

Though the report was brief, it contained numerous recommendations addressed to federal, state, and local governments, along with business. All of its recommendations were designed to dramatically increase the number of scientists and engineers entering the U.S. workforce, by:

- Building public support for making science, technology, engineering, and math improvement a national priority.

- Motivating more U.S. students and adults to pursue careers in science, technology, engineering, and mathematics.

- Upgrading K-12 math and science teaching to foster higher student achievement.

- Reforming immigration policies to attract and retain the "best and brightest" STEM students from around the world to study for advanced degrees and stay to work in the United States.

- Boosting and sustaining funding for basic research, especially in the physical sciences and engineering.[9]

The coalition of sponsoring business organizations embraced an ambitious (though never explained) quantitative goal for this education initiative. Indeed it printed this goal very prominently on the Report's front cover: *"Double the number of science, technology, engineering and mathematics graduates by 2015"*—that is, increase the number of bachelor's degrees awarded by U.S. colleges and universities in these fields by 100 percent within a decade of the report's 2005 publication date.

The Business Roundtable subsequently published an update in 2008, subtitled "Gaining Momentum, Losing Ground." It also maintains a website for its campaign to double the number of science, technology, engineering, and

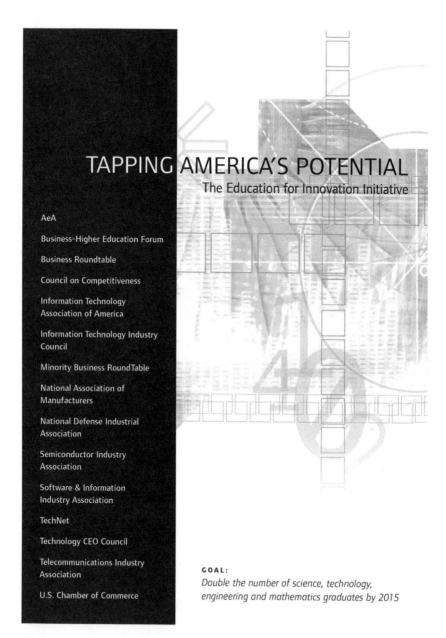

TAPPING AMERICA'S POTENTIAL
The Education for Innovation Initiative

AeA

Business-Higher Education Forum

Business Roundtable

Council on Competitiveness

Information Technology
Association of America

Information Technology Industry
Council

Minority Business RoundTable

National Association of
Manufacturers

National Defense Industrial
Association

Semiconductor Industry
Association

Software & Information
Industry Association

TechNet

Technology CEO Council

Telecommunications Industry
Association

U.S. Chamber of Commerce

GOAL:
*Double the number of science, technology,
engineering and mathematics graduates by 2015*

Figure 1.1.
Cover of 2005 report *Tapping America's Potential: The Education for Innovation
Initiative*.
Source: Washington, DC: Business Roundtable, 2005.

mathematics graduates by 2015, headlined by a revolving planet Earth with the caption "Are We Falling Behind?"[10]

Rising above the Gathering Storm ("Gathering Storm")

The third report was produced with unusual rapidity by the National Research Council (NRC), the executive arm of the National Academies.[11] An early draft was completed and circulated in October 2005, only a few months following the July publication of the TAP report. The final hard-copy volume, at 592 pages far longer and more detailed than the TAP report, was published by the National Academies Press in 2007.

The membership of this committee was dominated by current or former CEOs of large corporations and leading research universities. The report itself was both more detailed in its arguments and more restrained in its tone than *Tapping America's Potential*, although it did use evocative language such as the "Gathering Storm" metaphor in its title, "creeping crisis," "disturbing mosaic," and "the possibility that our lack of preparation will reduce the ability of the United States to compete in such a [globalizing] world."

The impetus for this report, and the process by which it was produced, both were quite unusual for the National Academies/National Research Council. The ad hoc committee was appointed in response to a request to the National Academies from four prominent members of Congress—all of whom were senior members of the relevant congressional committees.[12] They asked the Academies for prompt responses (within ten weeks) to the following questions:

- What are the top ten actions, in priority order, that federal policymakers could take to enhance the science and technology enterprise so that the United States can successfully compete, prosper, and be secure in the global community of the twenty-first century?

- What implementation strategy, with several concrete steps, could be used to implement each of those actions?

The National Research Council responded quickly by appointing a twenty-person ad hoc "Committee on Prospering in the Global Economy of the 21st Century." The committee was chaired by Norman R. Augustine, the respected former chairman and CEO of Lockheed Martin Corporation and former Under Secretary of the Army, and included five current or former chairmen or CEOs of very large corporations.[13] The other fifteen members included five current or former presidents of major research universities;[14]

six academic scientists or engineers, including three Nobelists;[15] three senior executives from National Laboratories and a leading pharmaceutical firm; a state superintendent of schools; and the founder of a foundation focused on strengthening education and research in science, mathematics, and engineering in Texas. (For a full listing of committee members, see table 1.2.)

In its report, the Gathering Storm committee honorably noted that the very short timetable required by the congressional request meant that it had been unable to undertake careful data collection and analyses of its own. Under these circumstances all of its recommendations had to be based upon the "consensus views and judgments of the committee members," bolstered by a necessarily rapidly prepared review of existing literature prepared by National Research Council (NRC) staff.[16] In short, this was hardly the kind of careful and deliberate analysis that would normally be expected from a National Research Council report; in the time available to the committee, it could not be. The draft report produced in this way was reviewed by thirty-seven experts invited by the National Research Council; these reviews too had to be completed in a highly expedited fashion.[17]

Given these conditions, it is perhaps not surprising that the *Gathering Storm* report's conclusions and recommendations were very close indeed to two other reports on the topic published only a few months earlier. It is also worth noting that there were nontrivial overlaps in the membership and staffing of the *Innovate America* and *Gathering Storm* reports: three participants served as members of both the "Principals Committee" for *Innovate America* and of the Gathering Storm committee; the chair of the Gathering Storm committee had co-chaired *Innovate America*'s advisory committee; and at least one person (David Attis) served as director, Policy Studies for *Innovate America* and as policy consultant for the *Gathering Storm* report.[18] It is impossible to know if these overlaps contributed to the similarity of these reports' recommendations. (For a side-by-side comparison of these report's key recommendations, see table 1.3.)

Conclusions of the NRC Report "Rising Above the Gathering Storm"

Any brief summary of the nearly 600-page report *Rising Above the Gathering Storm* would include the following key conclusions:

- Multiple trends variously described as a "quiet crisis," a "disturbing mosaic," or a "creeping crisis" suggest that the United States will be unable to compete in a globalizing world.

- These trends are "developing slowly but surely, each like a tile in a mosaic," due to failure to invest sufficiently over the long term in science and technology."[19]

Table 1.2. Membership of Committee on Prospering in the Global Economy of the Twenty-first Century, National Research Council

Norman R. Augustine
  (Chair), Retired Chairman and CEO, Lockheed Martin Corporation, Bethesda, Maryland
Craig R. Barrett
  Chairman of the Board, Intel Corporation, Chandler, Arizona
Gail Cassell
  Vice President, Scientific Affairs, and Distinguished Lilly Research Scholar for Infectious Diseases, Eli Lilly and Company, Indianapolis, Indiana
Steven Chu
  Director, E. O. Lawrence Berkeley National Laboratory, Berkeley, California
Robert M. Gates
  President, Texas A&M University, College Station, Texas
Nancy S. Grasmick
  Maryland State Superintendent of Schools, Baltimore, Maryland
Charles O. Holliday, Jr.
  Chairman of the Board and CEO, DuPont Company, Wilmington, Delaware
Shirley Ann Jackson
  President, Rensselaer Polytechnic Institute, Troy, New York
Anita K. Jones
  Lawrence R. Quarles Professor of Engineering and Applied Science, University of Virginia, Charlottesville, Virginia
Joshua Lederberg
  Sackler Foundation Scholar, Rockefeller University, New York, New York
Richard Levin
  President, Yale University, New Haven, Connecticut
C. D. (Dan) Mote, Jr.
  President, University of Maryland, College Park, Maryland
Cherry Murray
  Deputy Director for Science and Technology, Lawrence Livermore National Laboratory, Livermore, California
Peter O'Donnell, Jr.
  President, O'Donnell Foundation, Dallas, Texas
Lee R. Raymond
  Chairman and CEO, Exxon Mobil Corporation, Irving, Texas
Robert C. Richardson
  F. R. Newman Professor of Physics and Vice Provost for Research, Cornell University, Ithaca, New York
P. Roy Vagelos
  Retired Chairman and CEO, Merck, Whitehouse Station, New Jersey
Charles M. Vest
  President Emeritus, Massachusetts Institute of Technology, Cambridge, Massachusetts
George M. Whitesides
  Woodford L. & Ann A. Flowers University Professor, Harvard University, Cambridge, Massachusetts
Richard N. Zare
  Marguerite Blake Wilbur Professor in Natural Science, Stanford University, Stanford, California

Table 1.3. Parallel Report Recommendations

Innovate America (May 2005)	Tapping America's Potential (July 2005)	Rising Above the Gathering Storm (October 2005)
• More S&E undergrad/grad Incentivize universities Tax deductions for S&E scholarships	• More/better K–12 teachers	• More/better K–12 teachers
• 5000 new portable graduate fellowships	• More S&E undergrad/grad ∘ More scholarships & loan-forgiveness at all levels	• More S&E scholarships ∘ +25,000 4-yr undergraduate ∘ +5,000 3-yr graduate
• More S&E immigrants	• More S&E immigrants	• More S&E immigrants
• More research funding	• More research funding	• More research funding
• Permanent, restructured R&D tax credit		• Double R&D tax credit
• More Professional Science Master's programs		

• Human capital in science and engineering is the most essential long-term need of all economies, and the U.S. K-12 education system does "not seem able to produce enough students with the interest, motivation, knowledge, and skills they will need to compete and prosper in the emerging world."[20]

• Lower percentages of U.S. undergraduates pursue science and engineering degrees than in other countries. U.S. student interest in science and engineering has been waning, whereas students throughout much of the world see careers in science and engineering as the path to a better future.[21]

• The result is shortages of domestic scientists and engineers. The resulting gaps have been filled by foreign graduate scientists and engineers.

• Without intervention, "the nation could have difficulty meeting its need for scientists and engineers" if the number of foreign-born graduate students decreases as well.[22]

The tone of the *Gathering Storm* report is best conveyed by direct quotes from the report text itself. Under the heading "A Disturbing Mosaic," the NRC Committee offered the following summary of the views of its members:

> An educated, innovative, motivated workforce—human capital—is the most precious resource of any country in this new, flat world. Yet there is widespread concern about our K-12 science and mathematics education system, the foundation of that human capital in today's global economy.... Students in the United States are not keeping up with their counterparts in other countries.... After secondary school, fewer US students pursue science and engineering degrees than is the case of students in other countries.
>
> The domestic and world economies depend more and more on science and engineering. But our primary and secondary schools do not seem able to produce enough students with the interest, motivation, knowledge, and skills they will need to compete and prosper in the emerging world.[23] The United States ranks 16 of 17 nations in the proportion of 24-year-olds who earn degrees in natural sciences or engineering as opposed to other majors ... and 20 of 24 nations when looking at all 24-year-olds.... The number of bachelor's degrees awarded in the United States fluctuates greatly.... About 30% of students entering college in the United States (more than 95% of them US citizens or permanent residents) intend to major in science or engineering. That proportion has remained fairly constant over the past 20 years. However, undergraduate programs

in those disciplines report the lowest retention rates among all academic disciplines, and very few students transfer into these fields from others. Throughout the 1990s, fewer than half of undergraduate students who entered college intending to earn a science or engineering major completed a degree in one of those subjects. Undergraduates who opt out of those programs by switching majors are often among the most highly qualified college entrants, and they are disproportionately women and students of color. The implication is that potential science or engineering majors become discouraged well before they can join the workforce.[24]

There is still ample reason for concern about the future. A number of analysts expect to see a leveling off of the number of US-born students in graduate programs. If the number of foreign-born graduate students decreases as well, absent some substantive intervention, the nation could have difficulty meeting its need for scientists and engineers.[25]

Based on such views, the committee called for an ambitious array of new public policies to reverse these trends. These included:

• Make large government investments to "vastly improve" K-12 science and mathematics education, including annual recruitment of "10,000 science and mathematics teachers by awarding four-year scholarships and thereby educating 10 million minds," while also strengthening the skills of 250,000 current K-12 teachers through in-service training programs and economic incentives.

• Increase federal investment in long-term basic research by 10 percent each year for the next seven years.

• Increase the number and proportion of U.S. citizens who earn bachelor's and doctoral degrees in science, mathematics, and engineering by providing 25,000 new undergraduate scholarships and 5,000 new graduate fellowships each year.

• For foreign students who receive U.S. doctorates in science, technology, engineering, mathematics, or other fields of national need, extend visas by one year, and provide automatic work permits and expedited permanent residence visas if they are offered jobs by U.S.-based employers.

• At the same time institute a new skills-based, preferential immigration option.

As measured by direct impacts on specific legislation, the *Gathering Storm* report can be fairly described as one of the most politically influential reports ever produced by a National Research Council committee. Of course not all of its recommendations were enacted into law, but the four members of Congress who had requested this short-cycle report quickly transformed many of its recommendations into legislative language and introduced bills for congressional consideration, joining other pending legislative proposals on related topics.

After some delay caused by the vagaries of the legislative process, many of these recommendations were adopted by the U.S. Congress and signed by the president as part of the America COMPETES Act of 2007.[26] The Act authorized new or greatly expanded federal research funding for the National Science Foundation (NSF), the Department of Energy's (DOE) Office of Science, and the National Institute of Standards and Technology (NIST), with the research budgets for NSF and DOE to be doubled over ten years. At the K-12 level it authorized new grants to expand and improve teaching of science and mathematics and to better align it with postsecondary education, twenty-first-century workforce needs, and the Armed Forces. At the postgraduate level it authorized major expansions of NSF graduate research fellowship and traineeship programs, and mandated the NSF to facilitate development of professional science master's degree programs.[27]

The America COMPETES Act was "authorizing legislation," the first stage of the two-key U.S. legislative funding process—well-known for its complexity and apparent disorder, and often confusing to those not directly involved in it. Such "authorizations" can only *enable* the expenditures specified, but do not provide the needed funds. This requires passage of separate "appropriations" legislation, and this two-key legislative process had direct impact in this case. While many of the *Gathering Storm* report recommendations were indeed quickly incorporated into the authorizing language of the America COMPETES Act, the rapid pace with which the *Gathering Storm* recommendations were incorporated into law was then brought to a sudden halt by an unrelated legislative impasse that froze most of the annual appropriations bills during 2008.

During 2009, as the sharply negative consequences of the global financial crisis became apparent, appropriation bills were passed that incorporate funds for most of the expenditures authorized by the COMPETES Act, though only over the ensuing two-year period. These time-limited appropriations were part of the emergency economic stimulus legislation passed in 2009 under the American Recovery and Reinvestment Act (ARRA), which allocated substantial additional government expenditures during fiscal years 2009 and 2010 that proponents hoped would counteract rising unemployment and declining economic activity. Hence, though delayed by the appropriations impasse during 2008, the *TAP* and *Gathering Storm* reports can fairly be described as having achieved striking legislative success as part of the emergency economic stimulus. The additional funds appropriated were short-term only, however. It remains to be seen if the two reports' longer-term recommendations will be followed in the ensuing years.

Reprise: In 2010 the presidents of the three National Academies invited available members of the original Gathering Storm committee to provide a five-year update on the context and events since the original report was drafted in 2005. The update was drafted by the committee chair, Norman R. Augustine, and then refined by those members of the 2005 committee who were available.[28] Its conclusions were stated in brief and declarative language, building upon the "Gathering Storm" metaphor of the original report to forecast a storm of even greater power and destructiveness than had been anticipated in 2005:

> So where *does* America stand relative to its position of five years ago when the *Gathering Storm* report was prepared? The unanimous view of the committee members participating in the preparation of this report is that our nation's outlook has worsened. While progress has been made in certain areas—for example, launching of Advanced Research Projects Agency-Energy—the latitude to fix the problems being confronted has been severely diminished by the growth of the national debt over this period from $9 trillion to $13 trillion.
>
> Further, in spite of sometimes heroic efforts and occasional very bright spots, our overall public school system ... has shown little sign of improvement, particularly in mathematics and science. Finally, many other nations *have* been markedly progressing, thereby affecting America's relative ability to compete effectively for new factories, research laboratories, administrative centers—and *jobs*....
>
> The Gathering Storm Committee's overall conclusion is that in spite of the efforts of both those in government and the private sector, the outlook for America to compete for quality jobs has further deteriorated over the past five years.
>
> The Gathering Storm increasingly appears to be a Category 5.[29]

It was notable that the committee chose to build upon the metaphorical title of its original report by adding a vividly ominous metaphor to its subtitle—"Rapidly Approaching Category 5." This presumably refers to the catastrophic impacts of the most destructive category of hurricanes classified in the National Weather Service's Hurricane Wind Scale.[30] The destructive potential of a Category 5 hurricane is impressive: for example Hurricane Andrew, a Category 5 storm when it made landfall in South Florida in 1992, devastated the city of Homestead and surrounding areas and caused $48 billion in damage.[31]

Criticism of the *Gathering Storm* Report

Since publication of the *Gathering Storm* report in 2007, several empirical assessments of its data and analyses have been undertaken. Most have not been able to find credible evidence in support of the report's concerns about insufficiency in the numbers or quality of scientists and engineers being produced by U.S. higher education.

A 2007 paper by B. Lindsay Lowell and Hal Salzman examined the available empirical evidence about both the performance of U.S. K-12 students in science and mathematics, and the sufficiency of supply of graduates in these fields.[32] In brief, their conclusions are:

> Domestic and international trends suggest that U.S. schools show steady improvement in math and science, the U.S. is not at any particular disadvantage compared with most nations, and the supply of S&E-qualified graduates is large and ranks among the best internationally. Further, the number of undergraduates completing S&E studies has grown, and the number of S&E graduates remains high by historical standards.[33]

Their interpretation of the available data is that the *average* performance levels of U.S. K-12 students on international science and math exams is in the "moderate" range—neither high nor low. They find that this "moderate" average is a consequence not of weak or failing performance of those in the upper quartile, who are the most likely to pursue careers in these fields. To the contrary, their performance on these exams is strong. Instead, the U.S. average is pulled down to its "moderate" levels by the very poor performance at the lower end of the distribution.

Like most other quantitative analysts, Lowell and Salzman could find no credible evidence of insufficient supply of scientists and engineers in the U.S. workforce. To the contrary:

> Analysis of the flow of students up through the S&E pipeline, when it reaches the labor market, suggests the education system produces qualified graduates far in excess of demand: S&E occupations make up only about one-twentieth of all workers, and each year there are more than three times as many S&E four-year college graduates as S&E job openings.[34]

The authors do express support for efforts to improve average math and science education at the K-12 level, but do so out of concern for the needs of poorly performing students in the lower quartiles. They note however that given the above supply/demand situation, "such a strategy may not be the most efficient means of supplying the S&E workforce." They therefore conclude that:

> The available evidence points, first, to a need for targeted education policy, to focus on the populations in the lower portion of the performance distribution. Second, the seemingly more-than-adequate supply of qualified college graduates suggests a need for better understanding why the "demand side" fails to induce more graduates into the S&E workforce. Third, public and private investment should be balanced between domestic development of S&E workforce supply and global collaboration as a longer-term goal.[35]

A 2012 book by Yu Xie and Alexandra Killewald[36] also addresses the subject of the *Gathering Storm*, as indicated by its title: *Is American Science in Decline?*[37] It analyzes some of the same data studied by Lowell and Salzman, and offers a number of similar conclusions along with some differing interpretations.

Specifically, Xie and Killewald found no evidence of general shortages of scientists and engineers in the U.S. workforce. They agree that U.S. higher education routinely awards more degrees in science and engineering than can be employed in science and engineering occupations. However, they differ from Lowell and Salzman and from the National Science Board's *Science and Engineering Indicators* in arguing that the ratio of individuals with science or engineering degrees to those employed in science and engineering occupations is more like 2:1 than 3:1.[38] This difference is due to Xie and Killewald's decision to exclude the social sciences from the "science and engineering" category that is defined by the National Science Board and used by Lowell and Salzman. This also leads to higher completion rates among undergraduates who enter such science and engineering majors so defined, and higher rates of continuation into graduate degree programs among those who complete undergraduate degrees.

With respect to K-12 education, Xie and Killewald's analysis suggests that, when controlled for economic prosperity, the performance of U.S. K-12 students in internationally comparative studies such as TIMSS and PISA[39] is slightly below what might be expected in mathematics and slightly above in science. However, when compared to the two top-performing countries in TIMSS (Singapore and Hong Kong), U.S. performance "appears poor." In the PISA data, U.S. performance is somewhat lower in relative terms; in this study the top performers are "Chinese Taipei" (i.e., Taiwan) and Finland.

It should be noted that Xie and Killewald do not consider the possible distortions in such international comparisons of "average" performance that may result from the unusually high degree of inequality in U.S. K-12 education and the poor performance of its lowest student quartile, previously discussed. Nor do they stress the "city-state" nature of both Singapore and Hong Kong.

Overall, Xie and Killewald answer the question posed in their book's title with the conclusion that U.S. science (and engineering) is not in decline. Instead, U.S. science and engineering continue to strengthen and to maintain their positions as global leaders in many areas. They also find no sign of decline in the quality of U.S. K-12 education in science and mathematics; to the contrary they find it has been improving. At the same time, other countries that have lagged behind while the United States dominated global science and engineering have been catching up, and if their more rapid rates of improvement are sustained over many years, the United States may lose the dominant role it has held since World War II.[40]

Summary and Conclusions

The most credible evidence available is that at K-12 levels (or more specifically at middle and high school levels), the United States is by no means a world leader in *average* (or mean) student performance in science and mathematics. In comparison with other advanced countries, the *average* level of U.S. educational performance in these fields would best be described as "moderate" or even "mediocre."

These averages, however, disguise unusually high disparities in U.S. primary and secondary education. The performance in science and mathematics demonstrated by the top quartile of U.S. students is very strong, but the performance of the lowest quartile is very weak. In short, U.S. student performance in science and mathematics is unequal to a degree that may be unique among economically comparable countries.

Most countries in which average performance exceeds that of the United States show distributions of student performance that are much more homogeneous. In particular the lowest quartile of their students perform far better than does the lower quartile of U.S. students, although the performance of their top quartiles may be lower than that of the U.S. top quartile. There are many candidate explanations for these persistent educational inequalities in the United States: large economic inequalities, the heavy reliance upon local taxation to finance public schools, the absence of effective national, regional, or state mechanisms for raising the performance of the lowest-performing schools and students, and other factors as well.

Reduction of these very large differences between the top and bottom U.S. quartiles would be highly desirable in every respect. Competence in science and mathematics is essential to being adequately educated in the twenty-first century. The vexing challenge is in the "how": K-12 education in science and mathematics in the United States is, nearly uniquely among advanced industrialized countries, under the control of state and local governments, and indeed most of the (very large) financial resources devoted to this level of education come from state and local tax revenues. The governors of nearly all states have proclaimed that improvement of K-12 education is one of their highest priorities, and successive presidents and Congresses have devoted substantial political and financial capital to such efforts. State and federal initiatives, including "No Child Left Behind" and "Race to the Top," have provided large increases in financial support, but have proved to be politically controversial. Progress across states and local school districts has been highly uneven. While additional increases in federal funding and teacher training may be highly desirable if funds are available from constricted budgets, it would be wise to be skeptical that they will be able to cut or untie the Gordian Knot of U.S. K-12 education policy.

For the purposes of this volume, we must note that much recent discourse has tended to conflate the "moderate" or "mediocre" performance *on average* of U.S. K-12 students with the adequacy of the future U.S. science and engineering workforce. This is a sloppy argument, in part because it fails to recognize that the obviously large disparities in K-12 education mean that combining the very poor performance of the bottom quartile of U.S. students with the high performance of the upper quartile drives the overall averages downward, especially when compared to comparably advanced countries with greater homogeneity in their educational outcomes. Yet it is a fact that only a very small fraction of most countries' workforces are engaged in occupations that require high levels of science and mathematics—on the order of 5–10 percent—almost all of whom are drawn from the highest-performing student quartile. The poor performance of the bottom quartile is a very legitimate cause for real concern in terms of equality of opportunity and the overall education of the future citizenry and workforce, but it has rather less to say than might be supposed about the implications for the future U.S. science and engineering workforce.

Finally, there is a persistent tendency toward use of excessive language in public pronouncements about this subject. Phrases such as "the gathering storm" that is "now approaching Category 5" might suggest urgent and legitimate alarm about the serious, even catastrophic, damage that would result if current patterns were to continue."[41] They may also reflect however simply an astute assessment by sophisticated proponents as to just how difficult it is to capture the attention of U.S. political leaders about important education issues that really do need to be addressed. In the cacophony of U.S. political debates, screaming may be required if one wishes to be heard.

CHAPTER 2
No Shortage of Shortages

It is very difficult to make predictions, especially about the future.

—Danish saying

As noted in chapter 1, the three politically influential Washington reports released in 2005, *Innovate America*, *Tapping America's Potential*, and *Rising Above the Gathering Storm*, were by no means unique. To the contrary, they represent twenty-first-century renewals of fears expressed repeatedly during the second half of the twentieth century—hardy Washington perennials that have waxed and waned during nearly every decade since World War II. Indeed it is fair to say that many of the strengths and weaknesses inherent in the current U.S. science and engineering workforce, and of the higher education and research systems that are so central to it, are to substantial degrees lasting legacies of a half-century of governmental and institutional responses to episodes of such concerns that were usually accompanied by increasing and then declining demand for scientists and engineers.

To be sure, the rationales put forward for expanding the U.S. science and engineering workforce have shifted over time, and they also have differed from field to field, from industry to industry, and from proponent to proponent. What they have in common is a tone of alarm, and the frequent use of strong rhetorical language of national decline and even existential threat. The core argument is one of insufficiency, weakness, and national decay that is already well under way, or at least looming on the horizon. Examples abound.

In this view, the size, the quality, or the rate of growth in the U.S. science and engineering workforce simply is insufficient for the needs of the nation.

These unhappy realities represent critical threats to the national interest, to the nation's future prosperity, or even to national survival. Nothing good can come from trends in which the United States is increasingly falling behind its international competitors in its scientific and engineering activities.

The rhetorical power of such arguments is self-evident. Indeed, as can be seen from chapter 1, they have succeeded in capturing the attention of senior political, corporate, and education leaders. Unfortunately, they also have served to confuse serious thinking and to distort public policy.

One common element of such confusion is a tendency to address all levels of education together, from kindergarten through postdoc, in ways that pose profound challenges to credible generalization. This enormous educational terrain sprawls across categories of "students" who are as young as five and as old as thirty-five-plus years of age. It ranges from kindergarten entry, through elementary and middle schools, to high school performance and graduation rates, to college entry, to baccalaureate completion of a STEM major, to entry and completion of STEM master's and doctoral degrees, and even to completion of postdocs by recent STEM PhDs. The institutional sweep of such arguments is so broad as to include literally tens of thousands of schools, colleges, universities, and institutes. These differ widely in terms of type, size, and quality, making sensible generalization nearly impossible. Yet many such reports go even further, adding to this already unmanageable mix concerns about support for research and development (R&D), basic research output, and measures of productivity-enhancing technological innovation.

One of the most frequent and influential arguments of this type has been to conflate worries about the performance of the U.S. K-12 education system[1] over the long term with concerns about the adequacy of U.S. production of advanced degrees (especially doctorates) in science and engineering in the present-to-near term. The result is a deep logical gap in the argument, given the self-evident fact that rapid, even instantaneous, changes in the quality of science and mathematics education in elementary and middle schools cannot affect the number of advanced STEM degrees earned before the passage of at least ten to twenty years. Of course over the very long term these issues indeed are related to one another in some sense, but any effects of changes at the K-12 level could be seen only over a period of decades. In a shorter run of a few years, they are largely distinct and unconnected.

How then might one explain the common conflation of concerns about the quality of science and mathematics education at K-12 levels with argued insufficiencies in the production of advanced STEM degrees and in the size and growth rate of the U.S. science and engineering workforce? Why are concerns about U.S. K-12 education so often bundled with worries about workforce competitiveness in science and engineering? A number of possible motivations may underlie such claims.

The first and most encouraging interpretation would be that proponents have very long time horizons, and are urging actions to be taken in the short term that they fully understand could have effects only over the very long term.

A second interpretation would be that worries about U.S. national security or economic "competiveness" may provide a politically useful argument both for increased financial support for K-12 education and for additional government funding at the university level. Such arguments typically come from advocates for public education and for higher education.

A third interpretation is that claims that K-12 education levels are failing to provide sufficient preparation for STEM undergraduate study may be useful in deflecting criticism of those U.S. universities showing low retention and degree-completion rates among entering freshmen who had expressed prior interest in majoring in STEM fields. This used to be a common argument heard from leading research universities, although in 2012 the president of the Association of American Universities, the association of the leading research universities, offered the following forthright conclusion:

> [W]e now know that more than 50 percent of the students starting college with a stated desire to major in science or engineering drop out of those majors before graduating.
>
> We can no longer blame this problem entirely on the nation's high schools. A substantial body of research demonstrates conclusively that the problem is frequently caused by poor undergraduate teaching in physics, chemistry, biology, math, and engineering, particularly in the freshman and sophomore years.[2]

A fourth and more cynical interpretation is that claimed deficiencies at K-12 levels provide plausible arguments in support of large visa programs to admit "the best and the brightest" from abroad, at least as long as the K-12 system allegedly continues to produce inadequate numbers of graduates with strong skills in science and mathematics. The latter arguments can commonly be heard from corporate leaders seeking to globalize their workforce recruitment, and from immigration lawyers hoping to provide legal services for such visas.

While claims such as these are nearly universal in current debates, there is nothing new about them. Consider some of the rhetorical language used in *A Nation at Risk: The Imperative for Educational Reform*," a report to the nation produced in 1983 by a National Commission on Excellence in Education appointed by then-Secretary of Education Terrell Bell. Bell later wrote that his goal was to provoke a "Sputnik-type occurrence," and indeed the report concluded that the whole of the United States was "at risk" due to a crisis in the education that "undergirds American prosperity, security, and civility." Its schools and colleges were producing "a rising tide of mediocrity that threatens our very future as a Nation and a people."

An evaluative report produced later by Sandia National Laboratory at the request of the Secretary of Energy concluded that serious statistical errors underlay such conclusions. *A Nation at Risk* generated a great deal of attention and controversy that is far too voluminous to be discussed here; some additional detail is provided in appendix C.

A Nation at Risk deployed particularly memorable language linking the alleged failures of K-12 education to national security:

> If an unfriendly foreign power had attempted to impose on America the mediocre educational performance that exists today, we might well have viewed it as an act of war. As it stands, we have allowed this to happen to ourselves.... We have, in effect, been committing an act of unthinking, unilateral educational disarmament.[3]

While phrases such as "unthinking, unilateral educational disarmament" may be rhetorically excessive, concerns about U.S. science and math education often have been linked in more careful and modulated language to the most powerful political and economic developments of the past sixty years. These include the Cold War from 1948 to 1991; the 1950s political shock of Sputnik and Kennedy's 1961 promise of an American man on the moon within the decade; the Vietnam War and the rise of opposition to it; Nixon's declaration of a "war on cancer" in 1971; Reagan's defense buildup and Strategic Defense Initiative in the 1980s; the fall of Communism and the Soviet Union around 1990; the synchronized booms and busts of Internet, telecom, Y2K, and biotech in the late 1990s and early 2000s; the rising momentum of globalization and the offshore outsourcing of high-tech services from the 1990s to the present; and the effects of the deep global economic crisis that began around 2008.

For our purposes here, it is useful to consider five historical episodes or "rounds" of such concerns. Each round generated significant shifts in both supply of and demand for scientists and engineers. Each included active involvement by the U.S. government. While all had their own special historical and political characteristics and produced their own analytic literature and political rhetoric, they also shared a number of elements in common that can help to illuminate current and likely future episodes of similar concerns.

The first three of these historical rounds were closely tied to the Cold War, while the remaining two rounds occurred after the Cold War and had more to do with concerns about U.S. international competitiveness in the context of globalization. These two sets—the first about the Cold War, the second about globalization—produced similar outcomes, but differed notably in their politics and rhetoric. In part 1 of this chapter we first address the Cold War rounds, and then turn to the globalization rounds in part 2.

PART 1
Alarm, Boom, and Bust during the Cold War—Rounds 1, 2, and 3

By the end of World War II in 1945, the central importance of science and engineering capabilities for national security had been clearly established. It was widely understood that the resounding successes of the initially ill-prepared Allied military forces had depended in no small measure upon the prowess of the United States and its allies in wartime manufacturing on a gigantic scale, and in the success of their crash programs in research and development. As to the latter, it was clear to all that the outcomes had been greatly influenced by the Allies' successful development of radar, communication, and other powerful weapons technologies that had proved so decisive in conventional warfare. It was also evident that the use at Hiroshima and Nagasaki of the atomic weapons developed by the Manhattan Project had ended the Pacific war earlier and with far fewer Allied casualties than would have occurred otherwise.

ROUND 1 (1948–1957): Physical Sciences in the Cold War, and the Emergence of Biomedical Research

As the subsequent Cold War emerged and strengthened, the period after 1948 became a golden age of broad and enthusiastic public backing for expanded governmental support of scientific research and engineering. Moreover, science and engineering capacities were contributing significantly to the rapid growth of U.S. industries such as electronics and automobiles in the buoyant U.S. economy of the 1950s. The result was booming expansion in federal support for science and engineering, especially in the physical sciences.

National heroes included not only successful World War II generals such as Douglas MacArthur and Dwight Eisenhower (both contemplated presidential runs; the latter succeeded in 1952), but also nuclear physicists and "rocket scientists." David Kaiser, an MIT historian of physics, nicely captures the physicist-as-war-hero environment following the war:

> "Physical scientists are in vogue these days," announced a commentator in *Harper's* a few years after the war. "No dinner party is a success without at least one physicist." Physicists young and old— including those who had played no role in the wartime projects— found themselves "besieged with requests to speak before women's clubs" and "exhibited as lions at Washington tea-parties," reported a bemused senior physicist in 1950. Physicists' mundane travels became

draped with strange new fanfare. Police motorcades escorted twenty young physicists on their way to a private conference on Shelter Island, off the northern tip of Long Island, in 1947; a local booster sponsored a steak dinner en route for the startled guests of honor. B-25 bombers began to shuttle elite physicists-turned-government-advisors between Cambridge, Massachusetts and Washington, D.C. when civilian modes of transportation proved inconvenient.[4]

Large federal expenditures began to be directed toward research and development in the physical sciences, led by the Department of Defense and the Atomic Energy Commission. The justification was that research and development could provide real strategic advantage as the Cold War emerged between the United States and its erstwhile World War II ally, the USSR. The 1950s was a formative decade in which the Department of Defense and the Atomic Energy Commission became major funders of university-based research, while the National Science Foundation was created and began to receive significant appropriations. This was also the period during which the preexisting National Institute of Health was expanded to multiple Institutes and provided with substantial funds for the "extramural" research grants to universities and nongovernmental research institutes that later came to dominate its budget.

Among the various fields of science, physics especially claimed the mantle of victory in World War II through its contributions to decisive weapons systems such as radar, proximity fuses, and of course the atomic bomb. Although subsequent research suggests that this concentration upon physics may have been disproportionate given large contributions made by many other fields including chemistry, mathematics, and engineering, it became the conventional view at the time. By September 1953, the *New York Times* could declare with little disagreement that physicists were "responsible, among other things, for such of mankind's mixed blessings as refrigerators, television sets, and atomic bombs." To one leading student of the period, such claims were "wrong on all three counts, yet the formula seemed irresistible."[5]

The claim that physics as a field was the dominant intellectual basis of victory in World War II goes back to the onset of that war, when James B. Conant, then president of Harvard University (and a chemist by background), declared that this would be a "physicist's war rather than a chemist's…. For the present, at least, there appear to be more investigations of a physical nature than there are chemical military problems." The notion was embraced and energetically amplified by prominent physicists such as Henry A. Barton, director of the American Institute of Physics (AIP), the umbrella association of professional societies devoted to physics and physics teaching. Within forty days after Pearl Harbor, Barton began capitalizing on Conant's remark by publishing a series of AIP bulletins entitled "A Physicist's War."

Federal funding for science and engineering—and especially for physics—grew very rapidly in the decade following the 1945 end of the war. A widely shared belief in the critical strategic value of physics research meant that most of this funding came from the Department of Defense and the Atomic Energy Commission. Indeed, by 1949, no less than 96 percent of all federal funds for academic research in the physical sciences came from these two agencies, and their support for physics continued to increase even as the new National Science Foundation came on stream during the early 1950s as a civilian funding source for physics. By 1954, fully 98 percent of all federal funds for academic research in the physical sciences came from the Defense Department and the Atomic Energy Commission.[6]

As federal funding for physics increased rapidly, so too did both the supply and the demand for newly minted physics graduates, especially PhDs. Barton had expressed strong concerns that after the war the number of PhD physicists would decline. The subsequent reality proved otherwise. By 1948, the number of physics PhDs granted had risen 20 percent above the prewar peak. In the following year, 1949, the number of physics PhDs expanded by an additional 42 percent (see figure 2.1).

For at least some of the leaders of the key funding agencies focused on national security, this growth in PhD production actually was an articulated and important goal, rather than an unintended by-product of their funding for large physics research projects. Emanuel Piore, then-director of the Office of Naval Research's (ONR) Physical Sciences Division, described it as addressing the "deficit in technical people" after World War II "cost the country one to two graduate-school generations of scientists." Previously he noted that "Graduate students working part time are slave labor"—and hence the navy's best economic interests would be served by getting more and more of them on the ONR payroll.[7]

As the robust funding from defense and atomic energy agencies continued to expand rapidly, there were also additional federal sources of funding coming on stream. The early budgets of the National Science Foundation created in 1950 were quite modest, but they grew beginning in the mid-1950s until the mid-1960s, as may be seen in figure 2.2, which is expressed in "constant dollars," that is, adjusted for inflation. Funding then stagnated (or declined slightly in constant dollar terms) for two decades following the mid-1960s, until a sustained rise began in the mid-1980s and continued for the subsequent two decades. Then around 2006, the NSF budget was again reduced in constant dollar terms until around 2010, when it began to rise again slowly—here the trajectory is much confused by the sharp funding spike that is so apparent in figure 2.2 for FY2009 and 2010, a result of the

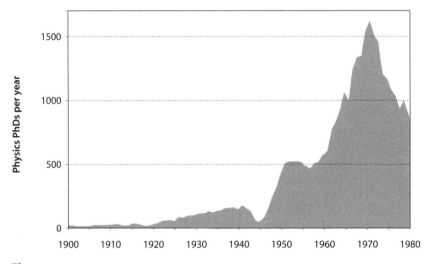

Figure 2.1.
Number of physics PhDs granted by U.S. institutions per year, 1900–1980.
Source: David Kaiser, "Booms, busts, and the world of ideas: Enrollment pressures and the challenge of specialization." *Osiris*, vol. 27 (2012): 276–302. Figure 2.

two-year pulse of "emergency" funding under the American Reconstruction and Reinvestment Act (ARRA), discussed elsewhere.

During this same period the first large funding streams began to flow to U.S. university researchers through the expanded extramural grants program of the already existing National Institutes of Health. Indeed, it was not only the physical sciences that produced national and international heroes in the postwar years. In 1952, Dr. Jonas Salk announced the first effective vaccine against the growing scourge of epidemic poliomyelitis (described as "easily the most frightening public health problem of the postwar era").[8] By 1955, following large-scale testing, the Salk vaccine was declared to be safe and effective—news often announced in screaming front-page headlines that was met with widespread national and international celebration.

Alarms and Cautions

Across the board, national funding for scientific research increased rapidly. The numbers of students pursuing undergraduate and graduate degrees in the sciences increased apace. Yet even with these new and rapidly increas-

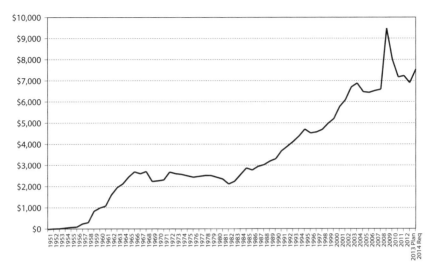

Figure 2.2.
History of total funding obligations, National Science Foundation, FY1951–FY2014, in millions of constant (2013) dollars.

Note: The surprisingly high spike in the NSF budget around 2009 is attributable to a special appropriation under the American Recovery and Reinvestment Act of 2009, abbreviated ARRA (Pub.L. 111-5), otherwise known as the economic stimulus act enacted following the financial crisis that began in 2008.

Source: Figure developed by author from data in NSF Obligations Funding History 1951–2012, including 2013 Plan and 2014 Request, National Science Foundation, Budget Office, June 25, 2013.

ing sources of government support for science and technology, early alarms began to be sounded about the United States falling behind in its science and engineering leadership. Such concerns were especially powerful with respect to America's Cold War adversary, the USSR, which had established its credibility in science and technology by testing its first atomic bomb in 1949, a much more powerful partial-fusion weapon in 1953, and a "true" fusion weapon in 1955 (the first U.S. test of a fusion weapon took place in 1952).

These increasing concerns about the United States "falling behind" the USSR led during the mid-1950s to several careful studies of the production of Soviet scientists and engineers. The first, sponsored by the National Science Foundation and the National Research Council and published in 1955, was conducted at the Russian Research Center at Harvard by Nicholas De-Witt, a Russian émigré from the Ukrainian city of Kharkov.[9] A second study by Alexander Korol of MIT's Center for International Studies, *Soviet Education for Science and Technology*, was published in 1957.

Both DeWitt and Korol scrupulously cautioned against over-interpretation of their reports' numerical estimates. They emphasized the vexing gaps in available data and the well-known Soviet propensities to exaggerate in such matters, and Korol warned against any "unwarranted implications" from directly comparing Soviet and U.S. rates of graduation for scientists and engineers, given major differences between the two countries' systems of higher education.[10] In particular, the Soviet numbers reflected large numbers of engineering graduates who worked in bureaucratic posts and never engaged in engineering work; the Soviet degrees were highly specialized into narrow subfields; quality standards appeared questionable; graduate numbers likely were manipulated to meet Soviet production quotas; and large fractions (about one-third of science and engineering enrollments in 1955) represented students in correspondence programs of doubtful quality.[11]

Subject to these caveats, DeWitt estimated that total undergraduate enrollments in the United States were three times greater than in the USSR. However, far larger fractions of Soviet than U.S. university students were concentrated in scientific or technical fields, so about 95,000 Soviet students were being graduated in scientific and technical fields each year (including the questionable correspondence school graduates) versus about 57,000 in the United States.[12] Subsequent assessments suggested that this "gap" might be expanding.

DeWitt and Korol's explicit warnings against direct numerical comparisons of U.S. and Soviet graduates in science and technical fields were mostly ignored by those citing their reports. The view that the USSR was graduating vastly more scientists and engineers took hold rapidly in pronouncements in the U.S. press and political discourse. Government officials, including CIA director Allen Dulles, began to speak frequently and forcefully of a "two-to-three times" ratio of science and engineering graduates, a gap that desperately needed to be closed. Some members of Congress called for a crash program to do so, and perhaps in response, in early 1956 President Eisenhower established a blue ribbon National Committee for the Development of Scientists and Engineers.

The leadership of the NSF and many leading scientists were more restrained about the claimed "scientific manpower gap." In its Sixth Annual Report for 1956, the NSF cited the importance of DeWitt's study (which it had partially financed), emphasizing that "there is great need to see the problem in its proper perspective." Dr. Lee DuBridge, one of the nation's most prominent physicists who at the time was president of Cal Tech and had previously served as science advisor to Presidents Truman and Eisenhower (and later to President Nixon), told the newly established National Committee for the Development of Scientists and Engineers that:

It is true that in Russia more men and women received degrees in science and engineering last year than in the United States. So what? Maybe that is because in the past 100 years they have so neglected their technical strength that they must now exert strenuous efforts to build it up. If this is true, then our rate of production should not be determined by their weakness—only by our own. Let us ask how many engineers we need to do our job, and not take over their figures for the numbers they require to do their job.[13]

In memoirs published in 1977, Dr. James Killian, president of MIT and the first chairman of the White House President's Science Advisory Committee (PSAC), described concerns about the argued scientific manpower gap as "patent nonsense."[14]

In his retrospective examination of the data on the Soviet "science and engineering workforce" produced by DeWitt, Kaiser notes a further complication that was not addressed at the time. The data on Soviet degree production in science and engineering included categories such as "agricultural scientists" and "health professionals," and the USSR graduated far larger numbers in these categories than did the United States. For the agricultural sciences, this Soviet focus was due in part to its later-discredited embrace of the agricultural ideas of Lysenko. Taking this and other data adjustments recommended by DeWitt into account, Kaiser estimated that

> the Soviets' much-ballyhooed lead shrank by a factor of ten, down to a mere 24%—a lead, moreover, that included a preponderance of students who earned their engineering diplomas armed almost exclusively with a textbook and a mailbox.
>
> Lies, damned lies, and statistics: the "manpower gap" proved no more threatening than the fictitious "missile gap" had been. And yet it still generated sufficient hot air—from high-ranking officials, widely read journalists, and influential physicists—to inflate American science classrooms far beyond any previous enrollment patterns.[15]

As suggested by this last comment, the general response in Washington was not the cautious one expressed by DuBridge and the NSF in 1956. To the contrary, there was palpable apprehension that the United States and its allies in the North Atlantic Treaty Organization (NATO)[16] were rapidly "falling behind" the USSR and its Warsaw Pact allies in human resources for science and engineering, fields viewed as absolutely critical for national security at a time of threatening Cold War.[17]

In September 1957, a NATO committee chaired by U.S. Senator Henry ("Scoop") Jackson (D-WA)[18] called for multiple science and mathematics educational initiatives to close the strategically threatening "gap." Jackson had for years been calling for such initiatives in the United States, but with little success because of general legislative stalemate and unrelated concerns about allowing a large federal role in education. In conjunction with his NATO committee, Jackson convened a stellar U.S. advisory committee of leading scientists, educators, and corporate leaders including Edward Teller, James Killian, Detlev Bronk, David Sarnoff, Maria Goeppert-Mayer, and Richard Courant. Together they urged the U.S. government to respond positively to the 1957 NATO report.[19] In this and other discussions of the period, the most often quoted number continued to be the "two-to-three times" cited by Dulles, derived from DeWitt's 1955 report but stripped of all of the caveats emphasized by its author.

Jackson and his advisory committee were of course highly influential figures in the United States, but Jackson's earlier failures to gain traction for such recommendations suggest that their efforts might have met the same fate—had it not been for the events that were to transpire only a few weeks after the NATO committee report was released in September 1957.

ROUND 2 (1957–1973): Sputnik, NASA, NDEA, Apollo, Project Hindsight, Vietnam, and the Mansfield Amendment

On October 4, 1957, the USSR stunned U.S. political elites by launching into Earth orbit a small 84-kilogram satellite called Sputnik 1. This launch had many effects in the United States, but perhaps the most visible was that the political shock it delivered was sufficient to galvanize actions that had long been promoted by Dulles and Jackson.

The astonishment expressed by politicians was strong, and yet the Sputnik launch actually was *not* a surprise at all—at least not for U.S. scientists, as we shall see directly. Moreover, there is considerable evidence that many in the U.S. government might actually have preferred to allow the USSR to be the first to launch an Earth orbiting satellite, though they may not have anticipated the domestic political firestorm that was to ensue. (See appendix A.)

Many U.S. scientists were very much aware that the launch of an orbiting scientific satellite had been agreed as an important and explicit goal of the International Geophysical Year (IGY),[20] and that both the USSR and the United States had active efforts under way to do so. The original idea for an International Geophysical Year, to include the launch of an orbiting scientific satellite, had been initiated in 1950 by an international group of atmospheric scientists, who correctly saw the developing capabilities of rocketry as a powerful instrument to advance their studies of the upper atmosphere.

The idea galvanized support from the International Council of Scientific Unions (ICSU),[21] from the U.S. National Academy of Sciences, and from Alan Waterman, director of the relatively new (and then still modestly funded) National Science Foundation, established in 1950.

In March 1954, the National Academy's IGY committee formally endorsed the idea of an attempt to launch a scientific satellite into orbit as part of the U.S. contribution to the IGY program.[22] Of course neither the National Academy of Sciences nor the National Science Foundation had the capacity for such a satellite launch; hence in spring 1955, their leaders (Alan Waterman for the NSF and Detlev Bronk, president of the National Academy of Sciences) approached the then-Secretary of Defense Charles Wilson with the idea that his department consider pursuing such a launch, in the interest of science and in the name of the National Academy of Sciences.[23]

The timing of this approach in support of launching a scientific satellite proved to be propitious from the perspective of the Department of Defense. The department and other U.S. intelligence agencies had become increasingly concerned at their repeated failure to accurately forecast the development of nuclear and ballistic capabilities in the Soviet Union. In particular, the 1953 Soviet test of a 400-kiloton nuclear weapon had been a shock, and the agencies had little or no information on Soviet nuclear ballistic missile programs. Indeed, in March 1954, President Eisenhower had told his Science Advisory Committee (of which both Waterman and Bronk were members) that he feared a surprise nuclear attack on U.S. cities, in part based on CIA estimates earlier that year that by 1957 the USSR would have a fleet of 500 of its turboprop nuclear bombers (the Tu-95, known as the "Bear"). Only a few weeks later, new alarms were raised when the Soviets once again surprised American observers by demonstrating a new Soviet full-jet bomber known as the M-4 or "Bison" at the public air show in celebration of May Day 1954.

The chair of the President's Science Advisory Committee, Lee DuBridge, responded to such concerns about inadequate intelligence by appointing a high-level Technology Capabilities Panel (TCP), chaired by James Killian, president of MIT. A subpanel of this group, known as "Project 3" and chaired by Edwin Land, inventor of the Polaroid camera, focused on improving technical means of surveillance. Clarence "Kelly" Johnson, one of the leading American aircraft designers and then director of Lockheed Corporation's Advanced Development Programs (generally known as the "Skunk Works"), proposed to this subpanel the rapid prototyping of a very long-range plane capable of maintaining surveillance from very high altitudes of Soviet nuclear and missile development and deployment. Moreover, he promised initial flights within eight months of a contract being signed. Land and Killian were impressed by the potential of such a program and obtained approval directly from Eisenhower, who assigned it to the CIA. The CIA designated this project with a deliberately innocuous name "Utility-2" —later shortened to "U-2."[24]

Hence the Technology Capabilities Panel had already stimulated development of one new means of "technical means of surveillance" when in February 1955 it issued its report, entitled "Meeting the Threat of Surprise Attack." It declared that "*We have the offensive advantage but are vulnerable to surprise attack....* Because of our vulnerability, the Soviets might be tempted to try an attack."[25]

This report in turn stimulated a further major technological effort, this time on the part of the U.S. Air Force, to pursue its long-standing interest in developing a reconnaissance system based on orbital satellites instead of planes—a project called "WS-117L." This was an idea that had originally been proposed by RAND and that had been percolating slowly within the air force.

Both of these top-secret surveillance projects—the "Utility-2" plane for the CIA and the "WS-117L" satellite for the air force—faced not only significant technological challenges, but also major uncertainties related to international law and diplomacy. Specifically, both were designed for surveillance over the USSR, the U-2 from an altitude of 70,000 feet and the WS-117L from an Earth orbit in space.

The legal and diplomatic problems the U-2 posed were fairly clear. The USSR long had fiercely asserted the sanctity of its sovereign airspace, even denying overflight rights to commercial flights as was the common practice elsewhere.[26] No matter how high it might fly in the atmosphere, the U-2 surely would be perceived by the USSR as infringing upon its sovereign airspace. While its 70,000-foot flight altitude might initially make it difficult to track by radar and to intercept, eventually the USSR would develop means to challenge U-2 overflights via improved radars and surface-to-air missiles (as, of course, did happen in 1960—also, symbolically enough, on May Day).

The legal and diplomatic issues for the WS-117L reconnaissance satellite program were far less clear. Since no one had yet orbited a satellite that passed over other nations' territories in "space" rather than "airspace," there was no precedent as to whether sovereign "airspace" was limited to a definable distance from the Earth's surface. Would an orbiting satellite "in space" violate the airspace of the sovereign "below" it? No one knew. There was no precedent, and no clarity.

As early as 1950, a RAND report had concluded that the USSR would be likely to view an orbiting U.S. surveillance satellite as a "highly anxiety-provoking" threat to its security, even as an act of "aggression" leading to equally aggressive Soviet responses. To mitigate such responses, the RAND study suggested a cautious strategy that first would establish an international principle of "freedom of space," parallel to the long-standing principle of "freedom of the high seas." Under such a principle a satellite orbiting over another country's territory would not infringe on sovereign airspace and hence would not require permission. The best way to do this would be to first launch a small scientific satellite (with none of the surveillance capabilities that might stimulate Soviet anxieties) into equatorial orbit, that is, well south of Soviet territory. Such a scientific satellite would necessarily cross the territories of many other countries, and if there were no complaints about infringement of sovereign airspace, the principle of "freedom of space" would be established.[27]

Given these strategic concerns, the Department of Defense found much of interest in the proposal to launch a purely scientific satellite as part of the International Geophysical Year. The question was how best to proceed. Both the navy and the army were developing rocket capabilities that might allow such a satellite launch and were competing with each other for priority. The navy's efforts (later known as Project Vanguard) were being conducted by the Naval Research Laboratory, which had a long-standing reputation for its basic scientific research, and the rocket boosters it was developing had been used in previous scientific research. In contrast, the army's efforts were based at its Redstone Arsenal, a center for military rocketry, and its candidate booster was a military rocket. Moreover, the Redstone Arsenal rocket program was directed by Wernher von Braun, famous (and for some infamous) for his leadership in developing the Nazi rocket weapon known as the V-2.

Clearly the navy's efforts had a distinct advantage in this respect, but the army's program had the best prospect of making the United States the first country to launch an orbiting scientific satellite. Indeed, the army leadership promised that it could launch such a satellite by January 1957 (see appendix A).[28] Though no clear directive appears to have been formally issued, a variety of actions suggested that on balance the decision was to emphasize the navy's efforts over those of the army, even if this slowed the overall U.S. effort. Indeed, there is some evidence that the army's satellite launch program was ordered into a holding pattern in 1956 (see appendix A).

> [S]peed was *not* the primary consideration; in the end the strongest civilian flavor in the project was more important. The administration was advised of the propagandistic value of being first into space. Of all these critical policy areas, however, the last had the lowest priority. For there were two ways the legal path could be cleared for reconnaissance satellites. One was if the United States got away with an initial small satellite—and had no one object to it. The other way was if the Soviet Union launched first.
>
> The second solution was less desirable, but it was not worth taking every measure to prevent.[29]

In the end, the "second solution"—that of the Soviets launching the first orbital satellite—did occur on October 4, 1957 with the successful launch of Sputnik 1, months before the first scheduled launch attempt of the navy's satellite program.

Political and Public Shock

While those in the defense and intelligence agencies were well aware of this "second solution," and indeed some may even have preferred it as more certain to establish the principle of "freedom of space," there appears to have been little awareness elsewhere among U.S. politicians, journalists, and the broader public before the successful Soviet satellite launch on October 4, 1957.[30] According to the chief historian of the National Aeronautics and Space Administration (NASA), an agency that was itself a product of the political reaction to Sputnik,

> [t]he only appropriate characterization that begins to capture the mood on 5 October involves the use of the word hysteria...."
>
> The launch of *Sputnik 1* had a "Pearl Harbor" effect on American public opinion. It was a shock, introducing the average citizen to the space age in a crisis setting. The event created an illusion of a technological gap and provided the impetus for increased spending for aerospace endeavors, technical and scientific educational programs, and the chartering of new federal agencies to manage air and space research and development.[31]

Sputnik's launch had deep impacts upon the members of the congressional leadership. Senator Jackson described the Soviet satellite launch as a "devastating blow" and urged President Eisenhower to declare a "week of shame and danger."[32] The reaction of Lyndon Johnson, the Democratic Majority Leader of the Senate, was described in memorable terms by the Chief Historian of NASA:

> On that same evening of 4 October, Senate Majority Leader Lyndon B. Johnson, Democrat-Texas, presided over one of his well-known barbecues at the LBJ Ranch in Texas. While at the gathering he heard the announcement of *Sputnik 1's* launch on the radio. He led his guests on a nighttime ramble about the ranch to the nearby Pedernales River, as he commonly did at such affairs, but Johnson's mind kept returning to the heavens as he pondered the Soviet triumph. He recollected, "Now, somehow, in some new way, the sky seemed almost alien. I also remember the profound shock of real-

izing that it might be possible for another nation to achieve technological superiority over this great country of ours."[33]

Less than a month later on November 3, the launch of Sputnik 2—six times heavier and carrying a dog named Laika—mobilized Lyndon Johnson into prompt action. He quickly organized urgent hearings of his Senate Defense Preparedness Subcommittee about the implications of the successful Soviet satellite launches. George E. Reedy, one of his staff members, noted that "the simple fact is that we can no longer consider the Russians to be behind us in technology. It took them four years to catch up to our atomic bomb and nine months to catch up to our hydrogen bomb. Now we are trying to catch up to their satellite."[34]

The Eisenhower administration responded by expediting the planned Project Vanguard launch, but this backfired badly when the launch failed and served if anything only to exacerbate the political shock. In the end both of the first two Vanguard launch attempts failed, and in highly public and notably spectacular fashion. The first attempt, on December 6, 1957, only two months after Sputnik 1, barely cleared the launch platform:

> During the ignition sequence, the rocket rose about three feet above the platform, shook briefly, and disintegrated in flames.

Lyndon Johnson's *cri de coeur* captured some of the emotion that by then had engulfed the issue: "How long, how long, oh God, how long will it take to catch up with Russia's two satellites?"[35]

The second Vanguard launch attempt took place only two months later, on February 5, 1958. It too proved to be a spectacular failure, exploding at an altitude of four miles above the launch platform.[36]

Meanwhile, the administration also had authorized the army's team of rocket scientists at the Redstone Arsenal led by Wernher von Braun to reactivate its satellite launch program, by then known as Explorer I. In the end the Explorer 1 was successfully and expeditiously launched into orbit on the army's Juno 1 booster rocket on January 31, 1958. This was less than four months after Sputnik 1, but the political shock had by then been delivered in a powerful way.

NASA, NDEA, and Apollo

Legislative responses to the Sputnik launches followed rapidly. Legislation signed in February 1958 created the Advanced Research Projects Agency (ARPA)[37] of the Department of Defense, later renamed DARPA (Defense Advanced Research Projects Agency). According to the history on the DARPA website, the agency

emerged in 1958 as part of a broad reaction to a singular event—
the launching by the Soviet Union of the Sputnik satellite on Oct.
4, 1957. While Sputnik itself—a mere 2-foot-diameter ball beeping
a radio signal—does not seem to be a particularly significant technological achievement, it had massive psychological and political
impact.[38]

In July 1958 the U.S. government's scattered and duplicative space and rocketry initiatives were reorganized into a new, legislatively created agency named
the National Aeronautics and Space Administration (NASA).

There was an explosion of concern too on the education front. An influential 1958 book by John Gunther, *Inside Russia Today*, argued that emphasis upon science and technology in the USSR's education system meant that
the average Soviet graduate from a secondary school had more science and
mathematics education than the average U.S. bachelor's degree recipient.
Moreover, concerns began to emerge that the USSR was graduating some
70,000 new engineers each year versus 30,000 in the United States.[39] In September 1958, only eleven months after the launch of Sputnik 1, Congress
passed and the president signed the new National Defense Education Act
(NDEA), precursors to which had languished for many years until the Sputnik launch provided crucial impetus.

> The potential for such a legislative breakthrough suddenly pre-
> sented by the launch of Sputnik was explicitly recognized by frus-
> trated proponents of Federal support for science and mathematics
> education. The Executive Director of the American Institute of Phys-
> ics, Elmer Hutchisson, did not mince his words: shortly after the
> Sputnik launch, he saw "an almost unprecedented opportunity to
> take advantage of the present public questioning concerning the
> quality of science instruction in our schools" and "the opportunity
> of influencing public opinion greatly." *Newsweek* quoted Hutchis-
> son as saying that the entire American way of life could well be
> "doomed to rapid extinction" unless the nation's scientific reserves
> were expanded quickly.[40]

Eisenhower, who reportedly had his doubts about the NDEA, signed it
only grudgingly in 1958 and described it as "short-term emergency legislation" to address the "Sputnik Crisis." Like many other "short-term" federal
programs before and after it, the NDEA continued in operation for fifteen
years, until 1973.

The financial support provided by the new NDEA legislation became
available during the same years that the numbers of Americans undertaking
undergraduate degrees across the board expanded substantially. Many have

assumed, reasonably enough, that the NDEA was itself the cause of these increases. However, the largest increases were seen in fields not supported by the NDEA, and an in-depth retrospective assessment conducted by the Institute for Defense Analyses concluded that it is difficult to draw causal relationships for most of the Act's provisions other than its Title IV, which provided substantially increased funding for NDEA graduate fellowships.[41]

The newly established agency NASA became operational in October 1958, and within one week its new administrator approved Project Mercury, designed to reassert U.S. supremacy over the USSR in space by putting a manned satellite into Earth orbit. Less than two years later, the Eisenhower administration approved initiation of NASA's Apollo program, a follow-on to Project Mercury designed to land an American on the moon. Nonetheless, the issue of national leadership in space became a prominent one in the 1960 presidential campaign. The Democratic Party presidential candidate, Senator John F. Kennedy, alleged frequently during his campaign that the Republican administration of Dwight Eisenhower not only had been inattentive as the USSR gained a substantial lead over the United States in space flight, but also had allowed the emergence of a direct threat to U.S. national security in the related "missile gap" in military weaponry. It was only after having won the very close 1960 election that Kennedy began in 1961 to acknowledge in private that the "missile gap" had been a myth. It was not until 1963, the year in which, sadly, he later was assassinated, that he so conceded in public.[42]

Less than three months after Kennedy's January 1961 inauguration, the USSR again trumped the accelerating U.S. space program Project Mercury by successfully launching the first human, Yuri Gagarin, into Earth orbit and returning him safely. Like the Sputnik 1 launch less than four years earlier, the Gagarin achievement provoked fears in some that the United States was falling behind the USSR in space and its related technologies of rocketry, ballistic missiles, computing, and guidance. Political pressures mounted on the new U.S. president to respond with a crash program to "catch up." Although Kennedy's presidential campaign had sharply criticized the Eisenhower administration for its alleged inattention to such matters, President Kennedy himself reportedly was initially cautious in response to such pressure. Before taking any action, he charged his new vice president, Lyndon Johnson, with a review of the feasibility of such an effort. It was perhaps unsurprising that the report by Johnson, a longtime supporter of U.S. competition with the USSR in space, supported an expanded space effort. Only six weeks later, on May 25, 1961, Kennedy gave a now-famous special address to a Joint Session of Congress. He called for an ambitious and expensive program for a manned moon landing by the end of the decade, which he framed in terms of competition with the Soviet Union:

Finally, if we are to win the battle that is now going on around the world between freedom and tyranny, the dramatic achievements in space which occurred in recent weeks should have made clear to us all, as did the Sputnik in 1957, the impact of this adventure on the minds of men everywhere, who are attempting to make a determination of which road they should take. Since early in my term, our efforts in space have been under review. With the advice of the Vice President, who is Chairman of the National Space Council, we have examined where we are strong and where we are not, where we may succeed and where we may not. Now it is time to take longer strides—time for a great new American enterprise—time for this nation to take a clearly leading role in space achievement, which in many ways may hold the key to our future on earth....

I therefore ask the Congress, above and beyond the increases I have earlier requested for space activities, to provide the funds which are needed to meet the following national goals: First, I believe that this nation should commit itself to achieving the goal, before this decade is out, of landing a man on the moon and returning him safely to the earth. No single space project in this period will be more impressive to mankind, or more important for the long-range exploration of space; and none will be so difficult or expensive to accomplish ...

Let it be clear ... that I am asking the Congress and the country to accept a firm commitment to a new course of action, a course which will last for many years and carry very heavy costs: 531 million dollars in fiscal '62—an estimated seven to nine billion dollars additional over the next five years. If we are to go only half way, or reduce our sights in the face of difficulty, in my judgment it would be better not to go at all....

This decision demands a major national commitment of scientific and technical manpower, materiel and facilities, and the possibility of their diversion from other important activities where they are already thinly spread.[43]

As he indicated, Kennedy was calling for major expenditures—the rough cost estimates that he provided in his 1961 speech would represent $50–65 billion in today's dollars, and the real costs ultimately proved to be far higher than Kennedy had imagined, or at least than he stated publicly (see later discussion). He also noted, correctly, that such an effort would demand "a major national commitment of scientific and technical manpower ... and the possibility of their diversion from other important activities where they are already thinly spread."

According to an official account published by NASA, Kennedy's proposals in May 1961 had "mesmerized" the public.[44] In due course Congress provided the support he requested, and with the substantial new funding thus made available a crash program commenced for a manned landing on the moon by the end of the decade. Yet by the summer of 1963, as the Mercury program was ending, prominent scientists and commentators began to ask questions about the race to the moon:

> [S]cientists began to see that the space program made distorting demands on skilled manpower, economic resources, and human determination. And they began to ask if it was really worth doing. Did we have to beat the Russians? Was this the most important scientific effort we could perform? Was NASA perhaps traveling too fast? The President himself seemed to have his doubts when he began to suggest joint space efforts with the Russians.[45]

Fortune magazine published a November 1963 article entitled, "Now It's an Agonizing Reappraisal of the Moon Race." The article asserted that the funding for *Apollo* was disproportionate to its potential payoffs, and would divert funds that otherwise would be available for unmanned space programs:

> Philip Abelson, Director of the Carnegie Institution's Geophysical Laboratory and editor of Science [magazine] ... had recently conducted an informal survey and found an overwhelming number of scientists against the manned lunar project. "I think very little in the way of enduring value is going to come out of putting man on the moon—two or three television spectaculars—and that's that," Abelson stated. "If there is no military value—people admit there isn't—and no scientific value—and no economic return, it will mean we would have put in a lot of engineering talent and research and wound up being the laughing stock of the world."[46]

In an address to the United Nations General Assembly on September 20, 1963, Kennedy did mention the possibility of a joint U.S.-USSR expedition to the moon in the interest of minimizing duplication of the large costs involved.[47] There were some tentative bilateral discussions of this idea, but these had not reached any conclusion when Kennedy was assassinated in Dallas two months later. It appears that the idea of such a joint effort was not pursued by his successor Lyndon Johnson.

The dramatic goal articulated by Kennedy in 1961 was achieved on July 20, 1969, when the *Apollo 11* mission successfully landed men on the moon's surface (and returned them safely to Earth on July 24, 1969). In the words of the NASA History Office, by this mission "the human race accomplished its single greatest technological achievement of all time when a human first set foot on another celestial body."[48] NASA also claimed major scientific advances from Apollo moon exploration: its Johnson Space Center compiled a list of the top ten scientific discoveries, primarily focused on the origins of the moon and the unlikelihood of any lunar life forms.[49]

For the purposes of this book, it suffices to understand that the Apollo program did indeed involve the "major national commitment of scientific and technical manpower, materiel and facilities" referred to by Kennedy. According to one estimate from NASA's Langley Research Center,

> [a]nswering President Kennedy's challenge and landing men on the moon by 1969 required the most sudden burst of technological creativity, and the largest commitment of resources ($24 billion), ever made by any nation in peacetime. At its peak, the Apollo program employed 400,000 Americans and required the support of over 20,000 industrial firms and universities.[50]

If we accept the estimate of $24 billion in 1960s dollars,[51] and adjust this for inflation using a variety of adjustment methods, Apollo program expenditures amounted to a minimum of $145 billion expressed in 2012 dollars, and likely far more.[52]

For our purposes here, however, the more important number is NASA's estimate that at its peak some 400,000 Americans were employed directly and indirectly in the Apollo program. This was indeed the "major national commitment of scientific and technical manpower" in Kennedy's speech, and gives substance to the concerns expressed by contemporary critics about the distortions it imposed upon the nation's skilled manpower and economic resources.

Ironically, it appears that the ultimate success of the *Apollo 11* moon landing in 1969 led to subsequent declines in political and public support for continued funding. During its peak appropriation years of 1966 and 1967, NASA funds for the Apollo program alone approached $3 billion annually (more than $20 billion in 2012 dollars), and constituted nearly 70 percent of the Agency's total appropriations. These very large government expenditures stimulated powerful demand for scientific and technical personnel. Yet by 1973 the appropriation for Apollo had declined by more than 95 percent from its peak levels only six years prior, to about $77 million. In that same year Apollo's share of NASA's budget had fallen to only 3 percent from nearly 70 percent in 1967, and NASA's total budget had itself declined substantially—by 40 percent in nominal terms from its 1967 level, and by more if inflation were factored in.[53] It is fair to say that rising demand for scientific and engineering personnel was first induced by heavy government financing of NASA's ambitious moon missions, and then, following the 1969 success of *Apollo 11*, declined almost as rapidly as it had increased.

Project Hindsight

Meanwhile, in another mission-oriented federal agency that had long been the source of heavy financial support for science and technology, an unrelated assessment was under way that ultimately would lead to further reductions in federal support for basic research in the physical sciences. As noted earlier, during the 1950s the Department of Defense (DoD) had become one of the largest supporters of university-based research in physics and related science disciplines. In July 1965 its director of Defense Research and Engineering requested that the military departments' assistant secretaries for research and development report on the payoff to the DoD of its large expenditures on research in science and technology. This report, titled "Project Hindsight," was completed in 1969.[54]

The report assessed the extent to which new DoD weapons systems were actually dependent upon the results of recent advances in science or technology; to estimate the fraction of such advances that resulted from DoD-financed research in science and technology; and to develop quantitative measures of return on investment in such research. In order to do so, a substantial study team was assembled from ad hoc groups of military and civilian DoD personnel; consulting behavioral and management scientists from Northwestern University and the MIT Sloan School of Management; contract support from economists at the Institute for Defense Analyses and the RAND Corporation; and volunteer support from five industrial/ management scientists.[55]

Project Hindsight was by no means a simplistic study. Its analyses were sophisticated and its conclusions nuanced. In general its 1969 final report tended toward the view that the most effective results of DoD support had not resulted from the department's support for what it termed "undirected basic research in science," but instead that the greatest payoff in terms of ideas leading to enhanced weapon systems has resulted from research in technology,"[56] primarily from applied research projects guided by mission-oriented agencies. Yet the latter actually could not replace basic or fundamental research:

> None of the findings … should be interpreted as a disavowal of the value of very fundamental research in science. It is hardly likely that the transistor could have been invented by people who sought a smaller more rugged electronic signal amplifier, but were unversed in wave mechanics or the theory of electrons in solids; that a search for new power sources could have led to the nucleus of the atom in the absence of work of Curie, Fermi and others; or that radio or telephone communications could have been invented without the research efforts of Hertz, Maxwell, Marconi and their fellows.[57]

Finally, the Project Hindsight report noted a different but important indirect result of research grants to universities: their role in educating the nation's scientists and engineers, of whom it estimated the DOD employed about one-quarter.[58]

Vietnam War and Opposition

The expansion of the Vietnam War and the political turmoil that followed also played a role—indeed two separate and distinct roles—in relation to the science and engineering workforce. First, as the size of the U.S. military commitment grew rapidly from 1965 and as domestic controversy about the war grew, increasing numbers of young men sought refuge from the military draft via the deferments readily available for both undergraduate and graduate students. These deferments became more difficult to obtain after 1971, but by then the U.S. war commitment was waning.[59] Second, there was growing political opposition to the war on the part of both students and faculty in U.S. universities, and in 1968–69 there had been much turmoil on university campuses in which opposition to Defense Department–funded research activities became a rallying cry.

In 1969, with little advance warning, Congress passed an amendment that placed limits on the kind of research funding that the Department of Defense could provide to universities. Known as the "Mansfield Amendment" after its sponsor, the influential Senate Majority Leader Mike Mansfield (D-MT), the measure required that all research—*including* basic research—funded by the Department of Defense must have a "direct and apparent relationship to a specific military function or operation." Mansfield himself was a prominent opponent of the war, and intended to diminish the role of the Department of Defense on university campuses. He had expected his amendment to reduce defense outlays for such research by $400 million, though the DoD later concluded that the amendment required only $10 million in cuts. DoD also sought (unsuccessfully) to have it repealed. When consulted, leaders of university-based science proved to be conflicted about the matter, given campus controversies about DoD-supported research. Most did not support repeal, but at the same time sought (also unsuccessfully) increases for the NSF budget to offset the expected DoD funding cuts.[60]

The Bust in the Natural Sciences and Engineering in the 1970s (but a concurrent boom in biomedical research funding from the "War on Cancer")

These three largely unrelated 1969 events—the spectacular success of NASA's Apollo moon landing; the Department of Defense Project Hindsight report;

and the Mansfield Amendment in Congress—all led to reductions in federal appropriations for basic research in the physical sciences. Project Hindsight and the Mansfield Amendment reflected doubts both inside the Defense Department and in the Congress about the wisdom of continued DoD funding for such basic research. The Mansfield Amendment itself was reversed after several years, but the trajectory of federal funding for such research during the 1970s was clearly downward.

Given the heavy role that had been played by defense funding since World War II and later by NASA, the physical sciences experienced constrained growth in their research and related funding. The effects were felt most immediately by recent degree recipients in these fields, many of whom expressed the view that they had been misled into pursuing physical science degrees and careers only to find dim prospects upon degree completion.[61]

As federal R&D funding for the physical sciences began to decline, the U.S. economy also experienced a relatively mild recession in 1969–70 that also had the effect of reducing the private sector's expenditures on R&D.[62] Both public and private support for the physical sciences waned, though there were some countervailing trends too. The 1971 National Cancer Act initiated large budget increases for the National Cancer Institute (NCI) and accelerated the rise of molecular biology, a field that had been attracting increasing attention from physicists.[63] Meanwhile the conquest of the dreaded polio had begun to attract increasing federal funding to the National Institutes of Health, and over the coming decades federal funding for biomedical research would greatly outstrip the growth in support for the physical sciences, as we shall see.

Overall, the earlier high rates of employment growth in academe for holders of science doctorates began to decline, with much of the growth that continued concentrated in the life sciences.[64] The result was evident to all, especially given the rapid growth in the student population during the 1960s expansion: a sharp deterioration in the early career prospects for new degree-holders in the physical sciences and engineering. The worries of the previous decade about looming insufficiencies in the numbers of physical scientists and engineers were rather suddenly transformed into concern about surplus numbers.

As disquiet spread during the early 1970s, some leading physical scientists produced gloomy projections of excess supply relative to likely demand, on the basis of which they began to advise cutbacks in the incentives for future students to pursue studies in science and engineering. One estimate on future supply and demand trends for PhD physicists and chemists, published by the distinguished MIT chemist Robert A. Alberty as part of a 1970 MIT symposium, suggested that equilibrium might require as much as a 50 percent reduction during the 1970s.[65]

There was, however, by no means a consensus about a need for what some called "birth control in science," with other leading scientists strongly counseling otherwise. The following year, in July 1971, the eminent American chemist Wallace R. Brode published a pessimistic article in the leading research journal *Science*.[66] Brode began by acknowledging that:

> Between 1968 and 1970, we appear to have moved from a deficit to a surplus of scientists and engineers. The surplus of scientists and engineers in the future will not be great, probably not more than 10 percent of the supply by 1983. But in a game of musical chairs in employment, any excess causes considerable displacement, especially among those entering the game.[67]

Brode clearly was concerned about the consequences for these new entrants to the science and engineering workforce, to which he returned later in the same article:

> Today the supply not only exceeds the demand, but is expanding at a greater rate than the demand. Perhaps the most disadvantaged of all are the new graduates, who have gone through or beyond college, even to postdoctoral training. They go out into the cold cruel world, only to find no jobs available, or else below the level of their training and ability.[68]

Brode did not mince his words about the difficulties of the science labor market in the early 1970s, but he saw this as an "exceptional" and "anomalous" situation that was certain to be reversed within only a 10–15 year period into a situation of "real shortage." He based this assessment upon his own rather simplistic demographic model of the forces determining new supply. In his view only a small percentage of the college age population (aged 18–22) had the ability and motivation to complete a degree in science or engineering. He estimated this to have averaged 3.8 percent, and considered it unlikely that it could ever exceed 4 percent.[69] This critical though weakly founded assumption, coupled with the fact that U.S. fertility and birth rates had begun to decline sharply during the 1960s, could only mean declining numbers of twenty-two-year-olds graduating with science and engineering degrees in the 1980s. Based on such assumptions, Brode argued strongly against scientists who urged that the rate of production of scientists and engineers be slowed to one more compatible with demand:

> In spite of an apparent surplus, we should encourage qualified students to major in science and engineering.... After 1983 the excess of scientists and engineers will taper off, and *by 1987 to the end of this century we are going to have a real shortage of scientists and engineers....* Judging from the number of births in the past decade, the demand will exceed the supply in the period from 1980 to 1990 ... we know there will be a shortage by 1990, if not earlier, which will last on into the next century.[70]

Brode was indeed a distinguished scientist, but his use of a simplistic demographic model to conclude that a shortage of scientists and engineers would occur by 1990 was not well-founded. Surprisingly enough, use of a similar model was to be repeated by policy analysts at the National Science Foundation some fifteen years later, as we shall see in Round 3. In both cases, the projected number of graduating scientists and engineers was determined solely by the number of U.S. births twenty-two years earlier, driven by the dubious assumption that a fixed percentage of each cohort was capable of such degrees. Both models allowed no impacts from trends in the economy, in occupational demand, in relative remuneration, or in other factors known to affect patterns of higher education and career choice.

ROUND 3: "War on Cancer"; Reagan Defense Buildup; and Federal Reports on Failing U.S. Education and Projections of "Shortfalls" in Scientists and Engineers (c. 1975–1995)

Like the beginning of the 1970s, the early years of the 1980s were also a period of economic downturn, this time in the form of a "double-dip" recession of considerably greater severity in 1981–82.[71] Yet there also was a sharp reversal in the downward trend of defense spending after the Reagan administration took office in January 1981. Over the ensuing eight-year period (and adjusting for inflation, i.e., in constant 2000 dollars), defense outlays rose by more than 41 percent and more than $116 billion, from $283 billion in fiscal year 1981 to almost $400 billion in fiscal year 1989.[72] These increased defense outlays in this period included major R&D efforts such as the Strategic Defense Initiative, famously dubbed "Star Wars" by the program's critics and many in the press.

Such expansions in defense R&D, combined with other federal R&D increases such as those for the War on Cancer, produced another surge in overall federal support for R&D during the 1980s, as can be seen in the lower panel of figure 2.3. As may be seen in the same figure, R&D outlays by industry were also rising substantially during this same decade.

"Shortfall" Studies at the NSF

By the mid-1980s, Brode's 1970s concerns about excessive numbers of scientists and engineers began to be replaced by echoes of the 1960s concerns about prospective shortages of scientists and engineers. In 1984, Erich Bloch retired from a long career at IBM[73] to become director of the National Science Foundation, and quickly established a small new advisory office for

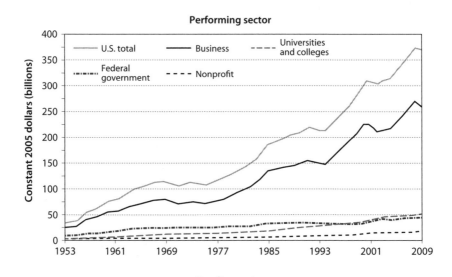

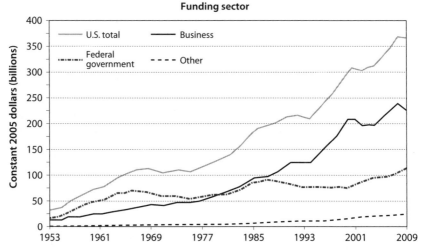

Figure 2.3.

U.S. R&D, by performing and funding sector, in billions of constant (2005)
dollars: 1953–2009.

Source: National Science Board, *Science and Engineering Indicators 2012*, fig. 4.4.

Policy Research and Analysis (PRA) led by Dr. Peter W. House. In 1985 this
group began an effort to develop its own projections of future supply of and
demand for scientists and engineers.

Over a period of about five years from 1986 to 1991, the PRA office at
NSF produced multiple drafts of reports that projected looming "shortfalls"
of U.S. scientists and engineers in the ensuing decades. Some of the projec-

tions were driven by simplistic demographic assumptions like those used fifteen years earlier by Brode. Based on its analyses, the PRA even presented specific shortfall numbers: initially it foresaw a rising shortfall that would cumulate to 692,000 by 2010; later it reduced this slightly to 675,000.

Within only a few years, however, it had become clear that these "short-fall" forecasts had proven quite wrong, and a 1992 congressional investigation of the NSF reports produced much embarrassment for the agency. The congressional investigators uncovered documentary evidence in the NSF files that NSF's own experts on the science and engineering labor force had sharply criticized the PRA forecasts as compromised by weak methodologies, dubious assumptions, and unjustifiable interpretations. They found too that the PRA reports were never approved formally for publication as official NSF reports, but that their findings of a looming "shortfall" of 675,000 scientists and engineers by 2010 were widely propagated via broadly circulated unofficial drafts, and by speeches, congressional testimony, and other presentations by NSF leaders.

What impact did such effects have? It is always difficult to prove causation in such circumstances, but the findings of these reports do appear to have been effective instruments in the hands of two politically active groups: those seeking large expansions in federal government support for science and engineering education, and those seeking new visas allowing ready access to international recruitment of scientists and engineers.

The first related directly to the NSF's request for substantial budget increases for its Education and Human Resources (EHR) programs. In his 1986 written testimony before the Senate Labor and Human Resources Committee, NSF director Bloch included the PRA projections of a looming shortfall of 692,000 new natural science and engineering degrees by 2010. He told the committee that "We are not attracting enough young people to science and engineering to assure an adequate supply for the future" (p. 4 of testimony). To respond to this insufficiency of supply, he requested an $89 million increase in NSF's FY1987 budget for science and engineering education.[74]

Figure 2.4 shows the trend of NSF education funding since its founding, in constant 2013 dollars to control for inflation. During the first decade of the NSF, when its overall budgets were quite small, education funding had represented large fractions of the total—generally on the order of 30–45 percent. Such funding increased sharply for a decade following the 1957 Sputnik launch, peaking in FY1968 at well over $700 million in 2013 dollars. There ensued a sharp and sustained decline, in which NSF's education funding plummeted by nearly 95 percent. At its low point in FY1983 it amounted to only $47 million in constant 2013 dollars, and constituted only 2 percent of the NSF total. These appropriations then began to recover from this low point, and by FY1987 had risen to nearly $200 million in constant dollars)—though this was still far below the level of just twenty years earlier.

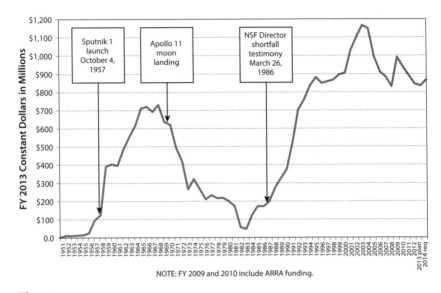

NOTE: FY 2009 and 2010 include ARRA funding.

Figure 2.4.
History of funding obligations for Education and Human Resources Directorate (EHR), National Science Foundation, , FY1951–FY2014, in millions of constant (2013) dollars.
Source: Figure developed by author from data in NSF Obligations Funding History 1951–2012, including 2013 Plan and 2014 Request, National Science Foundation, Budget Office, June 25, 2013.

Figure 2.4 shows that from FY1988 onward, the increases in congressional appropriations of NSF's Education and Human Resources (EHR) funding increased at exceptionally rapid rates—within a decade, in FY1997, the constant dollar figures were up to $857 million. These increases were far more rapid than those for the overall NSF budget, so by that same year the NSF's budget for education had risen to 19 percent of the total NSF budget. While it is difficult to be sure whether the NSF forecasts of large shortfalls of scientists and engineers contributed to this sharp funding turnaround, the timing is quite consistent with such an effect.

The second political effort to make effective use of the PRA shortfall studies was an industry-led campaign to establish new temporary visas for large numbers of skilled "specialty workers" from abroad. The new visas were justified on grounds that employers requiring scientists, engineers, computer and IT workers were facing debilitating "shortages" of qualified hires and thereby impeding their ability to compete internationally. Citations of the PRA findings of large "shortfalls" proved to be useful in establishing the credibility of these employers' claims. Congress responded to such claims by including a large, new visa category in the 1990 Immigration Act. This new visa (known by its legislative category "H-1B") later became the focus of a sustained industry lobbying campaign for expansion, again justified on grounds of international competitiveness and accompanied by a further round of workforce "shortfall" or "gap" studies produced not by NSF but instead by an industry trade association.

We now turn to two subsequent rounds of alarm/boom/bust that had nothing to do with Cold War competition but instead with economic competition in the context of globalization.

PART 2
Alarm, Boom, and Bust after the Cold War—Rounds 4 and 5

The Cold War ended in late 1991, with the formal dissolution of the USSR following the earlier collapse of the Warsaw Pact. In the United States this development naturally diminished political support for federal expenditures of a military/intelligence/strategic character, but also for those large-scale federal expenditures on science and engineering that had been premised on Cold War arguments. In 1993 a large federally financed research initiative in physics—the $12 billion Superconducting Super Collider—was cancelled by Congress, at least in part for such reasons.[75] The result was not a dramatic decline in federal R&D funding, but instead a more benign flattening of previously substantial rates of funding increases.

This slower growth in federal R&D funding in turn produced perceptions of a funding crisis in some fields of science and engineering, especially in physics and mathematics. However, another boom was to occur only 5–6 years later, this time driven less by federal funding and more by heavy private sector investments in emerging high-tech technologies (see discussion of Round 4). Coincidentally, the relatively flat funding for biomedical research during the 1990s led to a successful political campaign to double federal support for biomedical research during the five-year period 1998–2003 (Round 5).

ROUND 4: Concurrent Booms and Concurrent Busts in High Tech (1995–2005)

The late 1990s was an unusual period of nearly simultaneous boom times in several high-tech sectors: information technology (IT), Internet, telecommunications, and biotech. Of course there was some overlap among these industries, but each of these booms also had its own special drivers.

Information technology had been experiencing accelerating growth from the early 1980s, following invention of the microprocessor and the rapid spread of the first usable personal computers. Accelerating growth in IT turned to explosive boom in the early 1990s, with the release of the first credible version of Microsoft Windows.[76]

Meanwhile, what became the ubiquitous Internet of today had been developing slowly from its halting beginnings in the 1960s. Its origins were in Defense Department efforts to find ways to enable the government and military to communicate following a disastrous nuclear attack that destroyed other communication networks. For many years access was available only to government agencies and later to educational and research institutions. Its dramatic boom began in the mid-1990s, after access was first expanded to include business and commercial activities (the now-routine "dot-coms"), and the World Wide Web and early web browsers such as Mosaic became readily available.[77]

The telecom industry also boomed during the same period, led by upstart companies such as WorldCom, Nortel, and Global Crossing. These and other companies were able to attract huge amounts of capital to finance optical fiber infrastructure intended to meet the insatiable growth in demand that they and some financial analysts were forecasting. All three firms later imploded into bankruptcy, surrounded by accounting scandals and criminal charges against their managements.

Finally, the biotechnology industry also experienced a boom period in the 1990s, supported by a wave of investor enthusiasm for potential new and profitable therapies. Once again investment enthusiasm waned after the turn of the century, as it became clear that profitability was going to prove more difficult and slower than early investors had hoped.

In addition, in 1998 and 1999 there was an even shorter-term pulse of investment in the so-called "Y2K problem" (computing shorthand for "Year 2000") or "Millennium bug." The concern was about the internal "clocks" of legacy software elements that were still in wide use and that typically had been coded with only the last two digits of each year, for example, "89" instead of "1989," as a way to conserve then-scarce computer memory. No one knew what might happen to these software systems, already central to national and global economies, when the century ticked over from 1999 to 2000. As alarm spread during the late 1990s about the possible disastrous consequences, government and industry invested well over $100 billion in crash efforts to scrub millions of lines of computer code for the Y2K problems thought to lie therein.

Taken together, these simultaneous short-term booms and investment surges during the latter half of the 1990s produced sudden rapid growth in demand for IT workers; computer scientists; computer programmers; computer, electronic, and telecommunications engineers; molecular and biotech scientists; and more generally for scientists and engineers in many fields. By the middle of the year 2000 it was already evident that the Y2K problems actually had not produced any disasters, and the crisis funding was reduced sharply. Moreover, beginning in late 2000 the excesses were becoming apparent in the other sectors of simultaneous boom. As the end appeared of what later came to be called the "Internet bubble," the "telecom bubble," and the "biotech bubble," legions of small IT, Internet, telecom, and biotech firms that had been financed by the 1990s booms collapsed into bankruptcy. Others were acquired by larger firms and their staff sizes reduced. Large firms in these industries began to cut back drastically on their existing workforces, and to freeze new hiring.

In short, the simultaneous bubbles of the late 1990s in IT, telecom, software, and other high-tech fields all were punctured nearly simultaneously, producing one of the most powerful employment "busts" ever seen in science and technology fields. Tens of thousands of highly skilled employees suddenly found themselves unemployed, and many were stunned to discover they were unemployable in their own fields of specialization.

Lobbying for the H-1B

The H-1B visa for "specialty workers" was inserted into the 1990 immigration act after strong lobbying from industry employer groups. It provides employers with ready access to large numbers of temporary visas for foreign workers meeting specified criteria. The visas are valid for an initial three-year period, easily renewable for a second three-year period, and may be extended indefinitely beyond this six-year period if a petition for a permanent resident visa is pending.

While these visas often are described as limited to "highly skilled" personnel, the actual qualifications required are quite modest—the job must require a bachelor's or higher degree or its equivalent, and the visa recipient must have completed such a bachelor's or higher degree at an accredited U.S. college or university, hold an equivalent degree from a foreign institution, or have gained experience in the specialty that is equivalent to the completion of such a degree.[78]

As the concurrent high-tech booms emerged in the later 1990s for the IT, Internet, Y2K, telecommunications, and biotech industries, the same interest groups and their attorneys began to lobby strongly for expansion of these visas. With assistance from the American Immigration Lawyers Association (AILA), a new lobbying organization was created for this express purpose and located organizationally at the National Association of Manufacturers (NAM). Its initial name was *American Business for Legal Immigration (ABLI)*, later changed to *Compete America*. A former employee of the American

Immigration Lawyers Association, Ms. Jenifer Eisen, was appointed the ABLI's first director. By all reports financial support for ABLI came mainly from Microsoft, Cisco, Intel, and other IT companies. (See chapter 4 for more detailed discussion.)

The top lobbying priority for ABLI and its corporate supporters was to eliminate or greatly expand the numerical cap on H-1B visas, then set at 65,000 new visas per year. Since the H-1B visa allows temporary residence and work for three, six, or more years, the "stock" of H-1B visa-holders is much larger than the annual new inflow.

Supporters of H-1B expansion found their legislative champion in the then-chair of the Senate Immigration Subcommittee, Spencer Abraham (R-MI), a prominent libertarian conservative with close ties to the Cato Institute who was also a strong supporter of expanded immigration. In March 1998 Abraham introduced the "The American Competitiveness Act," designed to substantially increase the number of H-1B visas for foreign temporary workers, while also authorizing 20,000 scholarships for low-income students in math, engineering, and computer science and other more minor changes.

Energetic lobbying efforts ensued, led by numerous IT companies such as Microsoft and Intel, by the ABLI supported by these and other employers, and by the American Immigration Lawyers Association. Among the lobbyists retained to promote expansion of the H-1B was Jack Abramoff, who later became notorious following numerous scandals that surrounded his lobbying practice (see discussion that follows).

This intensive and well-funded lobbying effort achieved notable success after about three years. The renamed "American Competitiveness in the Twenty-first Century Act of 2000," now sponsored by Senator Orin Hatch (R-UT) with twenty-four co-sponsors including Abraham, was introduced in March 2000, passed the Senate and House of Representatives in October, and was signed two weeks later by President Clinton.[79] The Act tripled the number of H-1B visas from 65,000 to 195,000 each year for fiscal years 2001, 2002, and 2003.

In one of the interesting vignettes of U.S. politics, corporate lobbyists arranged for a provision in the Act that H-1B visas for higher education and nonprofit or government research would be exempt from any numerical limitations, in return for research universities' support for their efforts to triple the number of H-1B visas available to them.[80] (For a more detailed discussion of this lobbying episode, see chapter 4.)

The timing of this Act proved to be exquisitely poor. The multiple and simultaneous booms of the late 1990s in high-tech industries, which underlay the justifications for the expansion of H-1B visas, all ended with near-simultaneous busts just at the time that the first increases in H-1B visa numbers came into effect in late 2000.

ROUND 5: Rapid Doubling of NIH Budget, Followed by Crisis (1995–2008)

During the same years of the early 1990s, some of the leading spokespeople for U.S. science began to argue that there was a need to double federal funding for basic scientific research. Perhaps the most prominent was Dr. Leon Lederman, the distinguished physicist and Nobel laureate who had been an early supporter of the Superconducting Super Collider. As president-elect of the American Association for the Advancement of Science (AAAS), Lederman fielded what he called an "informal survey of key faculty in selected U.S. universities."[81] Based on the responses he received, he produced a personally authored report entitled *Science: The End of the Frontier*, published as a 21-page supplement to *Science* on January 11, 1991. The choice of this title represented a nice play on words on Vannevar Bush's influential 1945 report, *Science, the Endless Frontier*, which had laid the groundwork for federal financing of scientific research in universities and other non-governmental research institutions, and for creation of the National Science Foundation.

Lederman's report was written with passion and verve. It described the mood of the research community as one of "deep depression" that was "widespread, independent of institution, field and rank," and reflected "deep concern, discouragement, frustration, and even despair and resignation."[82] The gloom pervading the research community would surely be "communicated to undergraduates in an ever-widening ripple effect" and hence "in the coming years we can expect even fewer students to enter careers in science than do so now."[83] As a result, "American leadership in scientific research, as measured by published papers and Nobel prizes, is threatening to go the way of the automotive, tire, machine tool and consumer electronics industries."[84] The report argued passionately that "the daunting tasks in an ever more competitive world would require more scientists and engineers." And, interestingly, the evidence it cited for this claim included the National Science Foundation's PRA "shortfall" studies of the preceding years:

> Yet demographic projections—such as those cited by Richard Atkinson in his 1990 Presidential Address to the AAAS [based on the PRA "shortfall" studies, discussed earlier]—tell us that we are falling short of producing the required number of Ph.D. scientists and engineers by about 10,000 each year. Huge deficits in the number of technically trained personnel (estimated by some at up to 700,000) are expected in the first decade of the 21st Century.[85]

The report argued for, among other things, a doubling of federal support for science relative to that provided in the peak year of the "golden age of science" in 1968, adjusted for inflation. And beyond this first doubling, the report called for annual budgetary increases of 8–10 percent (which would imply a re-doubling every 7–8 years).[86]

The Lederman report received very substantial distribution. He honorably made clear that the report was his own and not an official AAAS document, although the AAAS decided to mail a copy to all of its 140,000 members as a supplement to the January 11, 1991 edition of *Science* magazine.[87] His conclusions were also conveyed to Washington policymakers at a meeting convened at the National Academy of Sciences on January 7, 1991.[88] As we shall see, the episode encapsulates many (if not all) of the elements central to this book: a well-intentioned and passionately written report produced by an eminent scientist that sounds the alarm about prospective deterioration of U.S. leadership in science and technology and its implications for U.S. competitiveness; invocation of forecasts of looming shortfalls of scientists and engineers in the decades ahead; a clarion call for federal funding increases to the levels of the peak years of the Apollo program; successful lobbying efforts contributing to booming but short-lived increases in federal research support for science, albeit not in the fields of greatest concern to the report author, as we shall see; and the emergence of serious funding crises once the years of rapid budget increases ended.

Dr. Lederman is a distinguished physicist noted for his contributions to the physical and mathematical sciences, yet his report called for the doubling of budgets in all scientific fields. Ironically his recommendations were embraced most effectively by advocates for the biomedical sciences, who focused primarily on doubling the budget of the National Institutes of Health (NIH), which provides only limited support to the physical and mathematical sciences. In part this was due to the biomedical sciences' better organization for advocacy. Supporters of NIH budget increases had already formed common cause via a lobbying alliance called *Research!America*. This lobbying group, originally created in 1989, was supported and jointly funded by a broad-based coalition of universities and medical schools; by many professional associations in the biomedical research field; and by large pharmaceutical and related companies in the healthcare sector. Perhaps more important, its goals were shared by energetic and committed voluntary associations seeking progress against major disease groups that are understandably of direct personal interest to many members of Congress.[89] In contrast, the array of groups supporting budget increases for the physical sciences (physics, chemistry, mathematics, and related fields, for which the NSF was a primary source of research funding) was far less potent in political terms.

Advocacy in support of doubling the budgets for scientific research percolated for several years, but initially achieved little because of the recession and budgetary stringencies of the early 1990s. By 1996, however, influential Republican members of Congress such as Senator Phil Gramm (R-TX) were developing proposals to double the budgets of NIH, NSF, and many other federal agencies over a ten-year period.[90] A resolution was introduced at the same time by Senator Connie Mack (R-FL) to double the NIH budget over five years, a proposal that was specifically welcomed by Senator Arlen Specter (R-PA), chair of the Appropriations subcommittee. Advocates for large increases in the NIH budget commented that in view of such political support they might raise their sights even higher than the 9 percent budget increase that had officially been requested by NIH.[91] Yet despite many words of support, in the end little was achieved in the 1997 appropriations round.

In its budget proposals for FY1999, however, the Clinton administration became supportive of large increases in a wide range of scientific research,[92] thereby joining with the mostly Republican congressional leaders who had been seeking such increases for several years before. Such bipartisan comity eroded later when it became clear that the Clinton administration's request for increased research budgets was to be financed by increased excise taxes on tobacco, projected to produce $10 billion by October 1998 and a further $55 billion by 2003.[93]

In the legislative efforts that ensued, the biomedical and health sciences proved to have many more strong supporters in Congress than did the natural sciences. Moreover, key congressional supporters of biomedical research were uniquely well-placed to influence the budgetary outcomes being sought. In particular, the Republican chairs of the Appropriations subcommittees with budget authority over the NIH in *both* the Senate (Arlen Specter) and the House of Representatives (John Porter, R-IL) were long-standing congressional advocates of increases in the NIH budget. Both had already formally aligned themselves in 1997 with the goal of doubling NIH's budget over five years.[94]

As the congressional elections approached in the fall of 1998, the appropriations process became mired in partisan politics that made it impossible to pass the majority of the appropriations bills. According to one informed account, this led to an "omnibus appropriation bill" (some 4,000 pages long) that combined eight of the stalled bills, the product of

> a marathon horse-trading session held behind closed doors ... biomedical research leaders are delighted with what they've seen so far ... a dream budget for NIH that should please even the agency's most ardent supporters. The omnibus bill ... boosts total NIH funding by 15%, putting it on track for doubling within 5 years. This matches the optimistic target set by biomedical advocates early this year—and is big enough to prevent destructive competition among disease interest groups.

The article reported that during this negotiation Senator Specter insisted on the near-15 percent increase for NIH, greatly exceeding the Clinton administration's request for an 8.7 percent raise and the House Appropriations subcommittee's proposal of 9.1 percent. The Clinton administration's proposal to use new tobacco taxes to finance these increases was blocked. Instead the Congress "contrived an 'emergency' that allowed it to break existing budget rules and use this year's $70 billion federal surplus to pay for a host of old and new projects."[95]

In a reprise the following year, supporters of doubling the NIH budget over five years scored another success when a last-minute agreement in November 1999 provided NIH with another 14.7 percent increase of $2.3 billion for its FY2000 budget.[96] That same year thirteen senators, including senior members of the relevant Appropriations subcommittee such as Kit Bond (R-MO) and Barbara Mikulski (D-MD), sent a letter to the Senate leadership calling for a parallel effort to double the budget of the National Science Foundation.[97]

As the 2000 presidential election approached, both the Republican and Democratic presidential candidates—George W. Bush and Al Gore—announced support of the NIH doubling goal. Gore but not Bush also registered support for doubling the NSF budget.[98] In December 2000, the coalition lobbying for rapid growth of the FY2001 NIH budget achieved another victory, when another appropriations compromise provided for a 14.2 percent increase ($2.5 billion) to the NIH budget.[99]

Following the contentious presidential election of November 2000, the first budget proposals for FY2002 submitted by the new administration of George W. Bush reflected its campaign promises: a 13.8 percent increase for the biomedical research supported by NIH, to $23.1 billion, thereby maintaining the trajectory for doubling the NIH budget over a five-year period, but little or no increases for natural science funders such as the NSF and the Department of Energy.

With the release of these Bush administration budget proposals, frustrated advocates for increased federal support of the natural sciences decided they needed to emulate the lobbying strategies that had been achieving such dramatic increases in the NIH budget. In May 2001, *Science* magazine reported that

> advocates for the physical sciences watched jealously as the biomedical lobby won a standing ovation for boosting the budget of the National Institutes of Health (NIH). Now a new influence group, the Alliance for Science and Technology Research in America (ASTRA), hopes to copy the biomedical lobby's tactics and pump up physical science budgets as well.[100]

It is impossible to be sure if this and other lobbying efforts had the desired effects, but in last-minute appropriations approved in December 2001, Congress not only increased the NIH budget by 14.7 percent or $3 billion, but also increased appropriations for the NSF by 8.7 percent (nearly $375 million) and those of the Department of Energy and other sources of physical science funding as well.[101]

In its next annual budget request, the Bush administration proposed and the Congress approved a final large $3.8 billion increase for NIH that completed the five-year doubling that had been promised. And the appropriations for NSF also began to show substantial growth.[102]

From Boom, to Hard Landing, to Funding Crisis

By the end of the five-year period from 1998 to 2003, the goal of doubling the NIH budget had been achieved. In current dollars, the NIH budget increased by more than 100 percent, from $13.6 billion to $27.3 billion.[103] Yet with completion of this remarkable achievement, robust budget growth for NIH suddenly ended. Following those five years of 14–15 percent annual budget increases, the NIH budget trajectory suddenly decelerated. In the ensuing years, excluding the two years of emergency ARRA funding, the NIH budget increased or decreased by only small percentages.

The result was what came to be known as the "hard landing" for biomedical research. To the surprise of some and to the consternation of all, the success rates for applications for the core "competing R01-equivalent" NIH research grants, which unsurprisingly had risen with the large funding increases, reached a peak around 2000 and then began to decline, gradually until 2003 and more rapidly thereafter (see figure 2.7). By 2004, success rates were lower than *before* the NIH budget began its sharp increase in 1998. The success rate had been nearly 25 percent in 1994 and rose smartly to 32 percent in 1999 and 2000, but by 2005 it had fallen to less than 23 percent. This was the case even though by 2005 the volume of funds available was well over twice as large as that in 1994. By 2006, this key success rate had declined even further and was approaching 20 percent. There ensued a temporary decline in the number of applications, which temporarily increased the success rates to about 22–23 percent from 2007 to 2010. But from then on the number of applications rose again, while the number of funded grants declined somewhat, resulting in further declines in success rate to about 18 percent in 2012.

Success with research grant proposals proved especially difficult for early-career researchers.[104] One of the concerns leading to support of the budget doubling that began in 1998 had been that the percentage of NIH awardees aged forty and younger had declined to less than 23 percent of the total. By 2005, the comparable percentage had declined to just over 15 percent.[105]

To some (modest) degree, these perplexing funding difficulties following so soon after truly massive budget increases were a function of inflation. Inflation rates of the period were not high, but costs for biomedical research rose more rapidly than general inflation. Figure 2.5 shows NIH appropriations

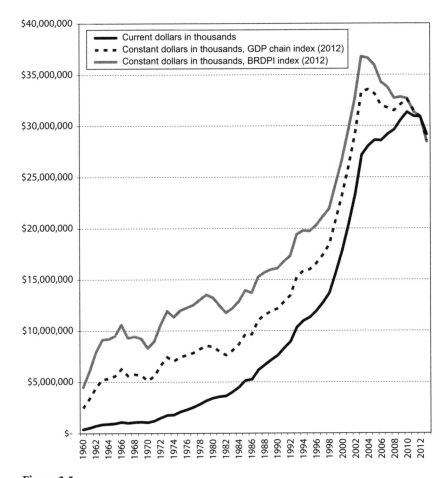

Figure 2.5.
NIH appropriations, FY1960–FY2013, in thousands of current dollars and constant (2012) dollars.
Source: Figure developed by author from data provided by Budget Office, NIH, June 25, 2013.

from 1960 to 2013 both in current dollars and in constant 2012 dollars, calculated on the basis of both of the inflation adjustments used by NIH (the GDP chain-type index[106] and the Biomedical Research and Development Price Index, or BRDPI).[107] Expressed this way, it is clear that the NIH budget over the 1998–2003 doubling period rose more rapidly than inflation as measured by both the GDP or BRDPI indices, but then failed to do so after the end of the doubling. Figure 2.6 illustrates another aspect of the NIH budget path by expressing it in constant 1960 dollars. It shows, unsurprisingly, that the NIH budget in constant dollars grew far less than in current

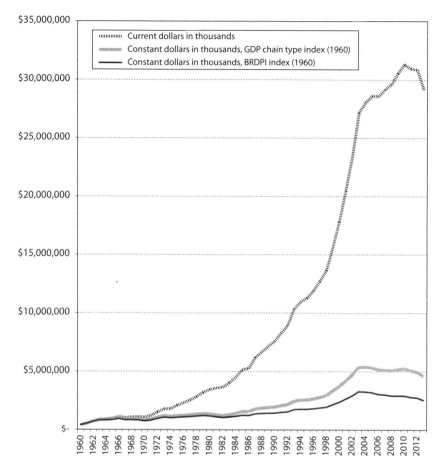

Figure 2.6.
NIH appropriations, FY1960–FY2013, in thousands of current dollars and constant (1960) dollars.
Source: Figure developed by author from data provided by Budget Office, NIH, June 25, 2013.

dollars, but also that even in constant dollars it increased between sixfold and tenfold depending on which inflation adjustment is preferred. There were still declines in constant 1960 dollars since the end of the budget doubling in 2003, but these appear more muted than when expressed in constant 2012 dollars as in figure 2.5.

The primary cause of the "hard landing," however, was not inflation, but instead the boom-bust character of the decade 1998–2008 —five fat years of booming annual increases (14–15 percent per year, and well over 10 percent annually adjusted for inflation) from 1998 to 2003, followed immediately

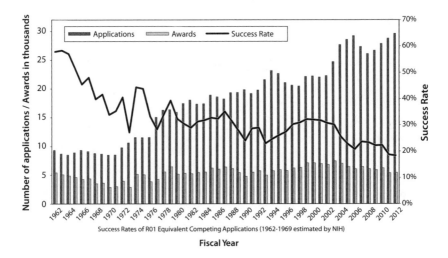

Figure 2.7.
Applications, awards, and success rates of NIH R01-equivalent competing applications, 1962–2012.
Source: Figure developed from data compiled by author from report.nih.gov, under "RePORT Catalogue," June 2013.

by five lean years of flat budgets (or declining budgets if adjusted for inflation).[108] Indeed, notwithstanding the massive increases from 1998 to 2003, by one estimate the real budget in 2008 was not much higher than it would have been had the far smaller annual NIH budget increases that had been common before 1998 been sustained from 1998 to 2008.[109, 110, 111]

With respect to the declining success rates of grant proposals to NIH, one important element was the rapid increase in the number of grant applications submitted. As shown in figure 2.7, as NIH grant budgets boomed between 1998 and 2003, the number of applications for the core "competing R01-equivalent" awards increased by nearly 50 percent, from about 20,000 in 1998 to more than 29,000 in 2006. It appears that more NIH funding begat many more NIH grant applicants,[112] while the number of applications per applicant also increased somewhat, presumably as applicants realized that the probability of success was declining for any particular proposal.[113]

The results proved damaging to biomedical research, research institutions, and researchers alike. We have already seen that the success rates for NIH grant applications declined to levels that were even lower than they had been before the budget doubling began. Increasing numbers of both established and new researchers found themselves unable to raise the external funds that were needed to support their laboratories and research activities. University deans were forced to provide "bridge funding" to researchers formerly supported by NIH grants. Even prominent scientists whose grant applications were not successful began to talk about closing down their labs, and fretted about the fate of the graduate and postdoc research assistants working for them. The language of distress shifted from "hard landing" to "funding crisis."

The powerful lobbying coalition that had successfully delivered the budget doubling between 1998 and 2003 tried its best to reverse the rot, but found little support from its erstwhile funders in Congress. The recessionary effects of the 2001 bust in the rapidly growing high-tech sectors imposed severe constraints on federal revenues, as did the Bush administration's tax cuts of 2001 and 2003. The military and other costs of responses to the attacks in New York and Washington of September 11, 2001 increased demand for federal resources in other domains. Many of the senior congressional leaders who had spearheaded the NIH doubling were no longer in office. And even members of Congress who were strong supporters of biomedical research did not seem responsive to pleas of a "funding crisis" from a research sector that had doubled its budget so rapidly only a few years earlier. The NIH budget was not increased by more than the rate of inflation, if that.

The 2005 National Academies report *Rising Above the Gathering Storm* (see more detailed discussion in chapter 1) seemed to break some of the budgetary logjam. Over a period of only a year or so, most of its recommendations became law as part of the 2007 America COMPETES Act, which authorized substantial increases in federal funding for science and engineering broadly defined. But the appropriations needed to actually provide additional funding were not forthcoming during 2008.

In March 2008, a group of presidents of leading research universities led by Harvard, UCLA, Duke, and Ohio State published a professionally produced pamphlet and website entitled *A Broken Pipeline? Flat Funding of the NIH Puts a Generation of Science at Risk.*[114] The report was released at a Washington press conference and followed by a Senate hearing convened by the late Senator Edward M. Kennedy (D-MA), then chair of the Senate Committee on Health, Education, Labor and Pensions (HELP). The report concluded that "An unprecedented five consecutive years of stagnant funding for the National Institutes of Health is putting America at risk—slowing the pace of medical advances, risking the future health of Americans, discouraging our best and brightest researchers, and threatening America's global leadership in biomedical research."[115] It cited some of the same statistics noted earlier, and emphasized the especially negative impacts of such funding problems upon the younger generation of researchers. The goal was a substantial increase in NIH's budget, on the order of a 6.7 percent increase for fiscal year 2009.[116] Yet once again it proved impossible to gain sufficient congressional support for such an increase.

And then, suddenly, came a short-term rescue from a most unlikely direction—the cataclysmic "meltdown" of the U.S. (and later global) financial markets that began in 2008. The crisis was initiated by collapse of an overheated U.S. housing market that had been driven by huge quantities of mortgage capital, coupled with what we now know to have been unethical and illegal behavior by financial institutions and lax government regulation of mortgage lenders and investment banks. The financial and economic crisis that resulted led to an unimaginable series of events, ultimately disgorging hundreds of billions of dollars in federal deficit spending for a variety of rescues and "economic stimulus" packages. One of these, the American Reconstruction and Reinvestment Act (ARRA) of 2009, included very large "emergency-style" funding for scientific research. Included in these funds were the first appropriations for the increased expenditures that had been authorized in 2007 under the America COMPETES Act.

Given the emergency atmosphere that prevailed, it is no surprise that the legislative process that led to these appropriations was rather chaotic. The Obama administrations had requested that an additional $4.4 billion in emergency funding be provided to NIH for expenditures over two years. When the smoke cleared, the NIH had received a far larger "stimulus"—a special appropriation of $10.4 billion over two years, to supplement its annual budget of about $31 billion. Again this short-term pulse of funding was heavily influenced by the insistence of Senator Arlen Specter, who considered the administration's $4.4 billion request to be insufficient and made it clear that his key vote would depend upon the addition of a further $6 billion. In the end his $6 billion demand was met. NSF, whose budget was not championed by Specter, received $3 billion of the same two-year stimulus funding.

The ARRA stimulus funds were specifically intended to be temporary emergency appropriations to be spent over only two fiscal years, and by 2010 increasing concerns were being heard as to what would happen when these funds ended—what was described as a looming "cliff" in biomedical research funding. Strong efforts were made to provide additional funds to the NIH, but budgetary stringencies limited the results and it remained unclear how the trajectory over the ensuing years would evolve.

Summary and Conclusions

Concerns about current and especially future shortages of scientists and engineers have been hardy Washington perennials since World War II. Most if not all have proved to be misleading due to the weaknesses of their methodologies and analyses. Indeed, the most sophisticated assessment of past efforts to forecast demand and supply of doctoral scientists and engineers, published by the National Research Council in 2000, concluded declaratively that "accurate forecasts have not been produced."[117] (For further discussion of the performance and challenges of such forecasting attempts, see chapter 8.)

Despite their misleading deficiencies, such claims have proven to be highly effective in political terms, capturing the attention of senior leaderships in government, business, and education. Because they have been politically influential they have often led to policy responses that in retrospect appear to have been confused or excessive, and that have stimulated successive rounds of alarm/boom/bust.

One source of confusion has been a tendency to conflate all levels of education together, from kindergarten to doctorate, and even beyond to the postdoc level. This leads to excessively gloomy prognostications of the future based upon only the "medium" or "mediocre" average performance (compared with other advanced industrialized countries) of U.S. primary and secondary school students on international science and math achievement tests. Such comparisons obscure two important realities: First, the domination of U.S. K-12 education by local and state levels of government, coupled with large differences in educational standards and financial resources across localities and states, leads to far larger disparities in K-12 education quality than do the equivalent public education systems of more homogeneous countries. The poor performance of the lowest quartiles of U.S. K-12 students pulls down the national performance averages, even while the top quartiles of U.S. students perform at high levels. Second, the science and engineering workforce represents only a very small fraction of the total workforce and is overwhelmingly drawn from the high-performing tiers of U.S. secondary school students.

We can identify five "rounds" of alarm/boom/bust since World War II. Most have lasted between ten and twenty years—initiated by the sounding of alarms about the insufficiency of the current or future science and engineering workforce, followed by governmental responses leading to booming growth in the number of scientists and engineers entering the workforce, followed by changes in circumstances that produce a bust in demand and chilly labor markets for new entrants. The first three of these cycles took place in the context of the Cold War, and indeed were heavily driven by its existential concerns. The final two occurred after the end of the Cold War, with the first driven by simultaneous but short-term booms in industries such as computing, IT, telecommunications, and biotech accompanied by successful lobbying by employers for ready access to global labor markets, the second by successful lobbying in support of large annual increases in research funding for the National Institutes of Health that proved ultimately to be unsustainable.

CHAPTER 3
Beliefs, Interests, Effects

How might we explain the repetitive cycles—over more than a half-century now—of *alarm/boom/bust* described in the preceding chapters?

In principle, the alarms that have initiated these cycles could have emerged on their own, based upon a broad-based national consensus reflecting widely shared concern about deterioration in the power, influence, and prosperity of the United States. In reality, however, such concerns often have been stimulated or encouraged by individuals and groups with strong interests at play, and especially so since the end of the Cold War. Who are these interested individuals and groups? What arguments do they put forth in support of their recommendations? And what outcomes have their efforts produced?

Interests during the Cold War

As we have seen in chapter 2, in the Cold War decades there actually was a consensus about the existential threat represented by the nuclear-armed USSR and its Warsaw Pact allies, accompanied by a broad-based belief drawn from World War II experience that U.S. leadership in science and technology would be critical to future national security. In fits and starts during the ensuing four decades, and especially during the 1950s and 1960s as the Cold War intensified, political and government figures involved in national security and leading members of Congress and the Executive branch raised concerns about the United States falling behind the USSR in science and technology capabilities essential to security.

Such concerns rose and fell. Their urgency increased each time the USSR surprised U.S. political leaders with its rapid advancement in scientific and engineering domains, first in nuclear weapons, then ballistic missiles, and then space technology. Attention subsequently faded over time as evidence emerged that the United States continued to be in the lead. Periods of increased concern tended to result in substantial pulses of federal funding for research and education in science and engineering fields, followed by flat or even declining support when concerns ebbed.

This pattern underlay at least three cycles of alarm/boom/bust during the Cold War period. The first example of such ebbs and flows (or perhaps we should call them flows and ebbs) was the rapid increase in support of physics from the Department of Defense and the Atomic Energy Commission, which dominated such support during the decade following World War II. As noted in chapter 2, such support was ebbing by the late 1950s, until an atmosphere of crisis engulfed Washington following the successful launch of the Sputnik satellites. The second, and much larger, example of flows and ebbs of federal funding for science and technology was the enormous investment in space and the moon landing program during the decade from around 1959 to 1969. A third cycle occurred after the rundown in defense expenditures following the Vietnam War was replaced by a large defense buildup during the 1980s, followed in turn by a subsequent decline after the end of the Cold War in the early 1990s. In all of these cases, increases in federal research funds increased demand for rising numbers of scientists and engineers, and graduation rates in these fields increased in response following the several years required for completion of such degrees. Then as federal funds subsequently ebbed, demand waned for previously highly sought-after scientists and engineers, and as new graduates in these fields experienced chilly labor markets, the numbers of students choosing to pursue study and careers in such fields also waned.

The process during the Cold War was relatively direct and straightforward. It is true that the scale of federal support for science and engineering might not have continued had it not been for the energetic advocacy of scientific groups such as the American Institute of Physics, the Cold War urgings of influential politicians such as Senators Jackson and Johnson, and the political shock administered to the political class by the successful launch by the USSR of Sputnik satellites. Yet the shared sense of national vulnerability during these four decades meant that the role of interest groups and their lobbyists seeking increased federal support for science and engineering, while real, was less significant than in the period following the end of the Cold War.

Interests after the Cold War

Since around 1990 there have been more cycles of alarm/boom/bust, involving a wider array of advocates than the mostly governmental and national security champions during the Cold War. Some government agencies and officials continued to be involved, but leadership in sounding new alarms about shortages of scientists and engineers shifted toward nongovernmental organizations and individuals, most prominently from four groups: employers, higher education, science funders, and immigration attorneys.

What arguments and evidence have they used in support of their claims of alarming shortages of scientists and engineers? How successful have they been in activating broader policy responses from the U.S. government? What key outcomes of these advocacy efforts can be discerned, and have these contributed to renewed cycles of alarm/boom/bust? And to what extent might these repeated alarm/boom/bust cycles—in perverse and quite unintended ways to be sure—turned out to be harmful to the public interest and to employers, education, and careers related to science and engineering?

General Decline of the United States

To some degree all of the postwar alarms—whether led by government or nongovernment voices—about the United States falling behind in science and engineering have reflected broad concerns about a *relative* decline in U.S. influence and leadership in both political and economic spheres—a late twentieth-century American version of the "end of empire" or "fin de siècle" alarms that were common in Europe a century earlier. If such general concerns about the relative decline of U.S. influence and leadership were widely shared, this could in part explain the ready embrace of such arguments by political leaders.

The United States emerged from World War II as a dominant economic and military power. Most of the other powers of the prewar period had experienced destructive warfare on their own territories, wholesale loss of life, or deep damage to their economies—in some cases all of the above. This was true of U.S. wartime allies such as the United Kingdom, France, and the USSR, as well as of wartime enemies such as Germany and Japan; both of the latter emerged from the war under Allied occupation and with much of their industrial infrastructure in ruins. Moreover, many scientists and engineers emigrated from war-damaged Europe to the United States, which was experiencing rapid postwar economic growth and large-scale expansion of its institutions of higher education and research.

The resulting sense of postwar military, political, and economic security for the United States did not last very long, however. In the context of the ensuing Cold War, alarms began to be sounded in the early 1950s about the rapid expansion of Soviet nuclear and missile capabilities. The Soviet Union had tested its first nuclear device in mid-1949, far earlier than Western intelligence agencies had anticipated, and tested its first partial thermonuclear device only four years later in 1953. This coincided with the period of McCarthyism in the United States, and indeed it was widely believed that Soviet espionage had contributed substantially to the rapid development of USSR nuclear capabilities. Though there probably will always be controversy about this matter, internal Soviet documents that were released following the collapse of the USSR suggest that such espionage activities may indeed have been quite real.[1]

Whatever the facts, by the mid-1950s McCarthyism was waning. Yet by then leading U.S. politicians such as then-senator John F. Kennedy also were pointing to what they termed the "missile gap," in which the USSR allegedly had established a dangerous lead in development and deployment of long-range nuclear missiles. As discussed in chapter 2, the political resonance of such claims was strengthened by the 1957 Soviet launch of the first orbiting satellites, to which many U.S. political leaders reacted with a self-reported sense of shock. They and other opinion leaders began to express concern that the United States was falling behind the Soviets due to insufficient expenditures not only on advanced weaponry, but also on science and mathematics education.

Such concerns waned in the late 1960s as the United States succeeded in its ambitious and expensive program initiated by President Kennedy to land the first person on the moon. But again, not for long: by the 1970s and 1980s, alarms that the United States was falling behind in high-tech fields were again being sounded. This time such concerns included anxieties about broader economic decline relative to international competitors, especially Japan. Rapid export-led economic growth in Japan was causing large U.S. trade deficits and declines in U.S. manufacturing. Japanese export growth was led by major Japanese business groups *(keiretsu)* consisting of interlocking corporations and banks that coordinated closely with and received strong support from the Japanese government. The success of these business groups contributed greatly to Japan's economic miracle that flowered most noticeably from the late 1960s to early 1990s. During this period, Japanese technology exporters rapidly and visibly established dominant positions in U.S. markets such as consumer electronics, and forcefully gained increasing shares of U.S. product markets in major U.S. industrial sectors such as automobiles.

During the same period, other developments including the costly U.S. intervention in Vietnam and Soviet expansionism in Asia (e.g., the Soviet military intervention in Afghanistan that began in 1979) and in Africa led also to concerns that the United States might once more be falling behind its Cold War adversaries in terms of military power. With the election of President Ronald Reagan in 1980, such concerns led to major expansions of U.S. government expenditures on military R&D and advanced weapons procurement.

The collapse of the USSR and the Warsaw Pact around 1990–91 removed that Cold War source of concern about falling behind in science and technology, but it was replaced by yet another source of anxiety—the accelerating departure of manufacturing from the United States to lower-cost venues beyond Japan. This phenomenon of "offshore outsourcing" was facilitated by improvements and declining costs of communication and transport, coupled with a rapid succession of significant U.S. government decisions to promote major "free trade" agreements. Indeed, some of the most prominent of these agreements were activated within a period of only seven years, mostly during the 1990s, including:

- the North America Free Trade Agreement (NAFTA) that came into force in 1994,

- the embrace of a new global trading regime under the World Trade Organization (WTO), agreed in 1995, and

- the decision to admit China into the WTO in 2001.

As U.S. manufacturing activities and their accompanying well-paid employment declined and U.S. trade deficits rose, the policy focus shifted to promotion of high-tech industries such as computing, IT, biotech, and telecommunications, all of which required substantial numbers of highly educated scientists and engineers. This contributed to arguments that highly trained scientists and engineers in these fields represented the critical "human capital" that was essential to the entire nation's economic future. These were "the best and the brightest," the leading edge of future "competitiveness" of the U.S. economy in a rapidly globalizing world.

Proponents claimed that due to its large, high-quality university system, the United States held a comparative advantage in its ability to produce large volumes of such "human capital," but was not investing sufficient public resources to do so. Others went further, arguing that there were actual or looming shortages of such human capital, and that those with degrees in science and engineering could expect robust demand for their skills in these industries and hence both remunerative and stable career paths.

For a while such claims turned out to be valid: all of these industries did experience powerful booms during the late 1990s, but they proved to be unsustainable as all of these booms turned to busts beginning around 2001. While they were under way, and actually long after they had waned, these high-tech booms raised their own alarms about whether U.S. educational systems were capable of providing the science and technology workforce for which industry advocates claimed there would continue to be strongly increasing need.

Looking at this U.S. experience from 1945 to the early decades of the twenty-first century from a long-term historical perspective, we can see some similarities to earlier episodes in which a previous global "hegemon" (a paramount power such as the British Empire in the nineteenth century) declined in relative terms. Earlier national and imperial powers did gradually lose their position of dominance not because they became weaker in an absolute sense, but rather because other countries closed the gap by achieving higher growth rates (economic, military, political) from lower initial levels. While the absolute power and prosperity of the hegemon continued to grow, its relative power could be seen by its leaders to be in decline, evoking expressions of alarm and calls for action to reverse the (relative) rot.

Case Studies of Industries Reporting Shortages

We turn now to discussion of the perspectives of those employers, industries, and educators that have been most active since the end of the Cold War in warning of the national dangers resulting from argued shortages of scientists and engineers.

In any such discussion, it is important to keep in mind the distinction between numerical "shortages" and the desire for lower cost employees. Employers of high-skill labor understandably would prefer to hire their employees at the lowest possible cost,[2] and even employers of low-skill labor often claim they face "shortages" when the evidence suggests large numbers of unemployed and underemployed workers with the necessary skills are available. The case of Fresno County, California, in which agricultural employers have long claimed labor shortages of low-skill farm workers even as county unemployment rates exceeded 10 percent for much of the past two decades and rose above 16 percent in 2010–12[3] is best understood as driven by the unwillingness of agricultural employers to offer pay and working conditions sufficient to attract U.S. workers.

What can be learned from the perspectives of the relatively small number of specific industries and groups that have warned of shortages in the supply of U.S. scientists and engineers, in contrast to a lack of such findings by objective researchers? Notwithstanding some conspiracy theories to the contrary, this apparent disconnect does not necessarily mean a willful rejection of convincing evidence.

Quantitative data on particular industries or occupations tend to be most readily available at the aggregate national level. It is entirely appropriate that analysts examine these data for credible empirical indicators of "shortages," which would include:

- unusually rapid increases in remuneration,

- unusually low or rapidly declining rates of unemployment and underemployment,

- unusually high or rapidly rising ratios of "open" relative to total positions in a field.

Such analyses do provide very important insights, but they also have limitations that warrant attention. Labor markets for scientists and engineers differ across several dimensions—industries, disciplines, locations, time periods, for example. Labor market surveys usually are not large enough to include sufficient numbers of respondents for specific years in relatively small geographic regions, specialized industries, and skill sets.

This means it is possible that no indicators of "shortages" would appear at the aggregate national level, even while employers recruiting science and engineering personnel could be experiencing challenging hiring conditions in particular locales, in particular specialized fields or industries, or in particular periods of rapid expansion. A charitable interpretation might therefore be that such employers could be generalizing their particular experiences to the national level—inappropriately to be sure, but nonetheless based on their own experiences—on the basis of recruiting difficulties in particular geographic locations (e.g., Silicon Valley), in particular technical subfields, or at particular times of expansive growth. The following discussion offers case studies of several industries that have expressed strong concern about generalized "shortages" of scientists and engineers.

Information Technology and Related Employment in Silicon Valley

During the late 1990s, employers in some then-booming industries such as IT, Internet, telecommunications, and biotech did face real difficulties in hiring the science and engineering staff they required for continued expansion. These industries have a lengthy history of strong boom and bust cycles, and as we have seen the dramatic booms of the late 1990s were quickly followed by dramatic busts beginning around 2001, followed by lower-amplitude booms and busts since then.

The aggregate boom/bust experiences in these industries were magnified in a few geographical areas in which such IT companies tended to "cluster." At the same time these industries are unusually mobile internationally, as evidenced by U.S. firms' rapid expansion of offshore outsourcing of IT functions to low-wage economies such as India.

The case of Silicon Valley in this period during the late 1990s offers a compelling case study. This was a period of exuberant boom, with rapid expansion fueled both by technical innovations and large inflows of funding from venture capital, equity markets, and other sources. In this frothy labor market, there were many reports of Silicon Valley employers actively seeking to "poach" key employees from other employers by offering large hiring bonuses and stock options, and numerous accounts of Silicon Valley scientists and engineers taking advantage of the situation to "job hop" their way to rapid increases in remuneration.

Given the heady booms under way at that time and in that particular place, such stories are entirely plausible. Unfortunately the situation changed rapidly and dramatically with the bust that followed a few years later, accompanied by numerous bankruptcies and large-scale layoffs of previously "hot" personnel.

Another part of this picture is that most of the science and engineering employment in Silicon Valley industries does not require professional certification or regulatory clearances, unlike the case of civil engineering for example. Hence many Silicon Valley employers may see their U.S. employees simply as part of an increasingly globalized labor market in which wages and benefits are subject to competitive pressures from low-wage countries.

Corporate executives from the IT industry have been prominent leaders of arguments asserting shortages of scientists and engineers. The most visible have been CEOs of large firms such as Microsoft (headquartered in its own IT cluster in the Seattle area), Intel, and Oracle, along with leaders of smaller but more numerous firms heavily concentrated in a few geographical regions and especially in the hothouse of Silicon Valley.

Domestic Petroleum and Gas Producers

In the domestic U.S. oil and gas industries, companies have experienced wide fluctuations in energy demand and pricing that led to rapid expansion and then contraction of their R&D workforces. As may be seen in figures 3.1 and 3.2, investment in energy R&D expanded sharply in conjunction with the rapid rise in real oil prices during the 1970s, driven by the Arab oil boycott and the subsequent Iranian Revolution and Iran-Iraq war. In response, private sector R&D rose to a peak of over $6 billion in 1980. Then, during the 1980s and 1990s, there was a sustained period of deregulation and declining real energy prices. One result was "significant downward pressure on the private sector's support for energy R&D throughout the 1980s and 1990s ... energy R&D bottomed out at approximately $1.8 billion in 1999," followed by some recovery in subsequent years to a level of $3.4 billion in 2005.[4]

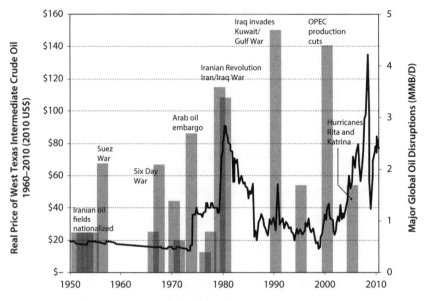

Figure 3.1.
Real price of U.S. crude oil per barrel, in constant (2010) dollars, and selected
major disruptions to world oil markets in millions of barrels per day (MMB/D).
Source: J. J. Dooley, "The Rise and Decline of U.S. Private Sector Investments in Energy R&D
since the Arab Oil Embargo of 1973." Report prepared for the U.S. Department of Energy
under Contract DE-AC05-76RL01830. Richland, WA, Pacific Northwest National Laboratory,
November 2010 (PNNL-19958), p. 5.

Energy industry recruitment of scientists and engineers tracked these
fluctuating commitments to R&D. Hiring was buoyant during the 1970s. It
then declined in the 1980s and 1990s, with some firms imposing hiring
freezes. This meant very good job and career prospects in this industry sector for entry-level scientists and engineers during the 1970s, followed by
deteriorating prospects for those at entry levels during the 1980s and 1990s.
After a multiyear lag, these weakening career paths then led to declining
interest in these fields among undergraduate and graduate students, and to
related contraction or closing of higher education programs in petroleum-
related science and engineering. The boom/bust cycle also meant that beginning in the 1980s the average age of these corporations' science and engineering workforces began to increase, with the large cohorts of new entrants
hired during the buoyant 1970s gradually aging in place while comparably
large new cohorts of younger job entrants were not hired.

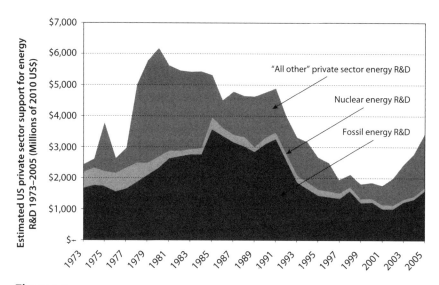

Figure 3.2.

U.S. private sector support for energy R&D, total, fossil and nuclear, in millions of constant (2010) dollars, 1973–2005.

Source: J. J. Dooley, "The Rise and Decline of U.S. Private Sector Investments in Energy R&D since the Arab Oil Embargo of 1973." Report prepared for the U.S. Department of Energy under Contract DE-AC05-76RL01830. Richland WA, Pacific Northwest National Laboratory, November 2010 (PNNL-19958), p. 7.

Early in the twenty-first century, real oil prices rose substantially from levels that had been stagnant or declining for the previous two decades. Meanwhile new technologies began to change the economics of Canadian tar sands and the Bakken shale oil deposits in northern North Dakota, Montana, and southern Saskatchewan. As firms again began to recruit new hires, they found that the numbers of graduating petroleum scientists and engineers was not increasing as rapidly as the suddenly rising demand. They also became aware that a substantial proportion of their existing science and engineering workforce, many of whom were recruited during the buoyant days of rising oil prices in the 1970s, had aged in place and were becoming eligible for retirement. The combination of suddenly increasing demand for petroleum scientists and engineers, stable or slowly rising supply of newly minted scientists and engineers in these fields, and rising numbers of prospective retirements led to rapid increases in the remuneration levels that needed to be offered to new recruits, and together led industry spokespersons to sound alarms about impending "shortages" of such personnel.

Nuclear Power

The U.S. domestic nuclear power industry is another case in which firms have recently been expressing concerns about an insufficiency in the numbers of scientists and engineers. This industry, like that of domestic petroleum, had been in the doldrums since U.S. enthusiasm for nuclear power plants began to wane in the mid-1970s as growth in demand for electricity declined below previously projected rates and utility company revenues declined. Another factor was the highly publicized accident at the Three Mile Island plant near Harrisburg, Pennsylvania in 1979, in which mechanical and operator failures resulted in a partial core meltdown. According to the International Atomic Energy Agency (IAEA) in an undated paper apparently written between 2005 and 2010,

> [t]owards the end of the 1970s the shrinking flow of nuclear power orders in the USA dried up completely, and it has not revived. The most obvious cause was the Three Mile Island accident on 28 March 1979, the first major accident at a civilian nuclear power station. The psychological effect on the population in the neighbourhood, and eventually throughout the Western world, was immense. So was the damage to the plant itself and to the reputation of the nuclear power industry.[5]

Public concern was heightened by the coincidental release, just twelve days before the Three Mile Island accident, of *The China Syndrome*, a popular Hollywood film about a nuclear plant meltdown starring Jane Fonda, Jack Lemmon, and Michael Douglas. The *coup de grace* for public opinion toward nuclear energy in the 1980s was provided seven years later by the nuclear catastrophe at the Chernobyl nuclear plant in Ukraine, which understandably greatly strengthened political opposition to approval and construction of new nuclear power plants in the United States. Had there been countervailing strategic or economic forces during this period, this opposition might have been less effective. But there were no international oil crises during the 1980s and 1990s, and indeed nuclear power became increasingly uneconomic as real prices for fossil fuels declined.

Then, after the turn of the twenty-first century, a new set of concerns became prominent and led to renewed interest in nuclear power. Evidence accumulated (though rejected by some) that burning of fossil fuels to generate electricity produced greenhouse gases that were contributing to climate change on a truly global scale, with unknown and likely unknowable but still disturbing implications. Meanwhile, for unrelated reasons oil prices also were on the rise. Together these trends caused some to conclude that nuclear power actually might be safer for the environment than carbon-based alternatives and also more plausibly economic over the long term.

Yet the previous two decades of little or no demand for nuclear scientists and engineers had gradually decimated the ranks of nuclear engineering departments and students in U.S. universities. As demand for such scientists and engineers suddenly began to rise, growth in their supply did not respond as quickly, leading to new claims about "shortages" of nuclear scientists and engineers.

Then, on March 11, 2011, there was another wrenching turnaround when the catastrophic magnitude 9.0 Tohoku earthquake off the Pacific coast of northern Japan triggered tsunami waves over 40 meters (130 feet) high that destroyed the large Fukushima Daichi nuclear power Plant on the northeastern coast. There were dramatic major meltdowns and gas explosions in three of its six reactors—evocative video images of which were televised around the world as the lead story for days and even weeks. Hundreds of thousands of residents threatened by releases of dangerous radiation had to evacuate the region. Though the long-run effects of this disaster cannot yet be assessed, there is little doubt that these events in Japan will have seriously compromised the public and political support that had been reemerging in the United States for nuclear power as an alternative to the environmentally problematic dependence upon fossil fuels. One likely outcome of this highly publicized catastrophe will be yet another reversal of trends in demand for nuclear scientists and engineers over the coming decade or more.

Aerospace

The aerospace industry is yet another example of an industry characterized by repeated booms and busts. The reasons are different from those driving booms and busts in information technology, petroleum engineering, and nuclear power. Aerospace firms are heavily dependent upon fluctuating demand from their primary customers, which depending upon the company in question are dominated either by commercial airlines or government agencies, both military and civilian. Moreover, the aerospace industry is characterized by additional complications related to national security.

On the civilian side, the primary customers for the aerospace industry's costly high-tech products are commercial passenger and freight airlines. These customers themselves have experienced cycles of strong growth and decline in demand for their services, due to fluctuations in economic growth that have well-known effects upon air travel and freight. In addition, the airline business is highly sensitive to major fluctuations in the costs of jet fuel, which have dramatic and often leveraged effects upon their profitability and hence their willingness and ability to invest in new aircraft produced by the aerospace industry.

Meanwhile, on the government side of the aerospace business, customers fall into two primary categories: defense and national security-related agencies, and civilian agencies such as NASA, the European Space Agency, and similar organizations devoted to space exploration.

As in the civil aircraft side of the aerospace industry, demand from U.S. defense and national security agencies has fluctuated greatly, driven by geopolitical developments (e.g., the collapse of the USSR and Warsaw Pact; the wars in Iraq and Afghanistan); by unpredictable events (e.g., the 9/11 attacks); and by domestic shifts in political control of the Congress and the executive branch. Similarly, the civilian government agencies that are leading customers of firms in the aerospace industry (such as NASA, Department of Homeland Security) have also experienced dramatic growth and decline in their procurement and contracting budgets.

Leading firms in the aerospace industry (Lockheed-Martin; Northrup-Grumman; Boeing's Defense, Space & Security division; Raytheon; and General Dynamics[6]) also depend upon large military procurement contracts. Unlike the IT and oil/gas industries, these firms understandably operate under requirements that substantial fractions of their science and engineering employees obtain appropriate security clearances. When such firms recruit from graduate science and engineering departments in U.S. universities that have enrolled high percentages of international students, they naturally face problems not because the U.S. universities have been educating insufficient total numbers of students in these fields, but instead because their new hires in such industries would be required to be "clearable," which usually is limited to U.S. citizens and legal permanent residents. The same would be true for many employees in U.S. government labs in these domains.

As noted earlier, many of the post–Cold War alarms about national shortages of scientists and engineers have been sounded by leaders of these industries. The recruitment difficulties they report may be real from their perspective; there is no reason to allege bad faith or dishonesty on their part. However, these real challenges may have been limited to particular periods of booming expansion, to particular disciplinary specializations that have gone in and out of favor, to the particular circumstances prevailing in geographical "clusters" such as Silicon Valley, or to the presence of large or small proportions of foreign students studying the disciplines from which they recruit in U.S. universities.

This means that the "shortage" claims put forward by such employers, their trade associations, and their lobbyists may have been overly generalized to the entire science and engineering labor market. The same can be said about some U.S. government agencies, where perspectives often are congruent with those in the industries to which they most closely relate. Those focused on military procurement naturally reflect concerns about the sufficiency of the relevant science and engineering workforce available for employment by military contractors (including the need for security clearances where required). The same is true for those focused on domestic petroleum and natural gas, and others focused on nuclear power.

Yet evidence put forward by organizations reflecting the perspectives of these particular industry sectors cannot be taken as representative of U.S. science and engineering in general. This naturally raises the question of what sources might provide more balanced and generalizable data and analysis, an important policy choice that will be addressed in chapter 8.

Is the Real "Shortage" in Research Funding?

Finally, federal agencies that provide substantial funds for research (most prominently the National Institutes of Health, the National Science Foundation, the Department of Energy, and the Department of Defense) have legitimate interest in ensuring that there are sufficient numbers of highly educated scientists and engineers to staff the research laboratories they manage and the external research activities they support. Agencies that are responsible for classified research and development also require that a least some of these employees be clearable.

The partnership between U.S. government research funders and research universities has been a productive one in terms of the stellar research that has been produced. With large-scale federal research grants providing financial support for both direct and indirect costs of high-quality research that would otherwise be well beyond their financial capacity, U.S. research universities have been predominant globally and continue to be so (for further discussion, see chapter 6). In some cases federal research funding provides substantial portions of the budgets of university science and engineering departments, although academic administrators often argue that calculations of the "indirect" costs for such research grants fail to cover substantial parts of the true costs of such research. In return, federal funding agencies have succeeded in enabling achievements in basic science that might not have been possible had the same funds been deployed to research laboratories outside of universities.

Meanwhile university faculty members with strong research interests have been financed by federal agencies to pursue important scientific issues that fascinate and energize them. They in turn are judged by their peers (for tenure, promotion, scientific reputation, awards) primarily on the basis of their research output.

For many researchers this system seems well-designed, even optimal. Science is a "calling" akin to those of religion, or music, or environmental protection. They are truly committed, animated by their love of science to devote much of their intellect and energy to its fascinating challenges. They believe that the public should see science for what it is: not only intellectually challenging and a source of wonder and beauty, but also a true public "good" that produces knowledge and technologies that serve the interests of all, both domestic and global.

How can there be too much of such a public good, provided by brilliant and committed researchers who devote their lives to its pursuit? Why do they find themselves facing vexing problems of obtaining the funds needed for such research?

From this perspective, committed scientists and engineers of outstanding quality are forced to spend an undue fraction of their time writing proposals for research funding. Large fractions of the proposals they develop and submit are judged scientifically meritorious but then cannot be funded—a huge waste of time by those who could otherwise be rapidly pushing back the frontiers of scientific understanding. From this perspective, the real shortage in the system is not in the number of scientists and engineers, but instead in the failure of the federal government to provide enough funding to support research by the growing numbers of committed scientists who seek it. Because research funding also provides support for their graduate students and postdocs, they sometimes join forces with advocates claiming shortages of scientists and engineers in the broader workforce.

This view of a paucity of federal funding is consistent with the data available from the largest federal science funding agency, the National Institutes of Health (NIH). As may be seen in figure 2.7, less than one in five research proposals received by NIH in FY2012 could be supported with the funds available; this means that four out of five of the proposals submitted to the NIH (a process that is labor-intensive and time-consuming) were not funded. Figure 2.7 also shows that the "success rate" (the solid line graph) declined by one-third during the first decade of the twenty-first century; it exceeded 30 percent in 2001, and was less than 20 percent in 2011. As may be seen in figure 2.7, the underlying dynamics of this decline in success rate were driven by relative constancy in the number of funded grants during this period—on the order of 6,500—while the number of proposals submitted increased greatly—from about 22,000 in 2001 to about 28,000 annually between 2005 and 2010.

Hence, one obvious solution is for the federal government to appropriate more funds for research. Indeed, as we have seen in chapter 2, this case for additional funding for NIH was made successfully during the 1990s, leading to the budget doubling from $13.6 billion to $27.3 billion between 1998 and 2003. Success rates for proposals submitted to NIH did increase over this period of expansionary budgets, although when NIH budget appropriations subsequently ceased their rapid growth after 2003 the success rates declined to levels even lower than those before the doubling began.

The experience of the NIH budget doubling from 1998 to 2003, and the funding crisis that ensued when rapid budgetary growth ended, provides useful insights into some of the structural problems that can now be seen in the research funding system described earlier. One of the problems is that, whereas this system does indeed produce truly world-class research, it also produces additional PhDs and postdocs as a kind of by-product. In simplified form, the structure works like this: a funding increase for research provides financial support for an increased number of graduate students and postdocs, who have come to be the preferred bench-level research workers for federally funded research in universities. After a multiyear lag or delay, increased numbers of young scientists who are able to find eligible research positions seek to emulate their mentors by submitting research proposals to NIH. (For a fuller discussion, see chapter 6.)

There are other important structural factors operating as well. Research funding provides valuable sources of revenue for research universities, and university administrators therefore often encourage their faculty to seek more research funding and provide incentives to those who do so. Scientists seeking NIH grants realize that the success rate for grant applications is low, and therefore they submit more proposals to increase their chances of succeeding. Together these forces lead to large expansions in the numbers of grant applicants and grant applications that, over time, produce declining success rates—even though available research funding has increased substantially.

Summary and Conclusions

A number of episodes of national alarm about argued insufficiencies in the supply of scientists and engineers initially were driven by concerns about national security in the context of the existential threats of the Cold War. These concerns were initiated primarily by political leaders and executive branch agencies focused on national security, joined later by advocates for substantial government investment in the improvement of U.S. education in science and mathematics from kindergarten to graduate school.

More recent episodes of national alarm about "shortages" of skilled workers emanated less from government and national security sectors and more from nongovernmental organizations and activists primarily in four domains:

- corporate employers, especially those in computer-related and information technology industries, and more recently from petroleum producers responding to the rise in oil prices;

- some in Congress and in government agencies that finance research and education in science and engineering, who either have been convinced by such private sector claims of shortages or who see them as useful means to argue for increased budgets for their agencies or favored programs;

• some in K-12 and higher education, for whom shortage claims provide compelling rationales for increased public investments in their work;

• immigration lawyers and advocates, for whom shortage claims are useful arguments in support of their interest in expanded numbers of permanent and/or temporary visas for skilled personnel from abroad.

From the perspective of many university-based researchers, the real shortage is not one of scientific and engineering professionals, but of federal funding for basic research in universities. This system will be discussed in greater detail in chapter 6, but we turn first to the activities and successes of those pointing to current or prospective shortages of scientists and engineers.

CHAPTER 4
The Influence of Employer and Other Interest Groups

I have said it thrice:
What I tell you three times is true.

—Lewis Carroll, *The Hunting of the Snark* (1874)

As discussed in chapter 2, the several rounds of alarms about insufficiency of the U.S. STEM workforce were dominated by the national security concerns of the Cold War, and led to a variety of actions by government to educate more U.S. scientists and engineers. Since the Cold War ended around 1990, similar alarms have focused more on the economic significance of the STEM workforce and have often been led by employers seeking to recruit more foreign scientists and engineers, although usually conjoined with parallel calls for improved education in the United States.

The strategy of expanding international recruitment that has been a focus of recent advocacy was not really an option before around 1990, since large pools of foreign scientists and engineers (other than in the USSR and Warsaw Pact) were not available, and certainly not in countries with far lower wage levels than the United States. However, just as the national security imperative was waning during the 1990s, there also was dramatic expansion in the number of scientists and engineers available in low-wage countries such as India and China.

Many labor market experts have expressed concerns that excessive use of this foreign worker option might backfire by diverting U.S. citizens away from science and engineering majors and careers; however, most large U.S. employers with global operations have not seemed troubled by such concerns. Indeed these claims that domestic STEM workforce "shortages" require expansive access to international recruitment have been dominated by U.S. employers, their employer associations, and their lobbyists, joined by lobbyists for the immigration bar and higher education. Together these interest groups have produced, financed, mobilized, or led most of the visible and influential reports, congressional testimony, press stories, op-eds, press releases, and other means to communicate their views.

Very substantial funds to support such efforts have been provided by employers and their associations. Much of the advocacy involved has emphasized the need to greatly expand employer access to temporary visas such as the H-1B (discussed briefly in chapter 2) that allow employers to easily hire STEM workers from abroad. In political terms, current and impending STEM workforce shortages promoted by such interest groups have been demonstrably successful in achieving their legislative goals.

Some critics see these shortage claims as deliberate and cynical falsehoods put forward by employers who know they are not supported by credible empirical evidence. This is possible, but not proven. A more benign view is that those making such claims are honestly reflecting experiences of U.S. corporate employers and universities in recruiting graduate students and employees in particular fields and subfields, especially those discussed in the case studies in chapter 3. If true, their persistent claims of widespread and generalized STEM workforce shortages would indeed be excessive generalizations based on limited cases, but they may nonetheless reflect honest belief rather than deliberate falsehood. This also is possible, but not proven.

This chapter addresses recent studies arguing that current or looming shortages in the STEM workforce require expanded numbers of temporary visas (such as the H-1B) for skilled migrants, and clarifies the oft-garbled requirements of such visas. It compares such findings with the growing body of independent research conducted by prominent labor economists and other experts on these labor markets. It concludes by considering the lobbying activities undertaken to support shortage claims and expanded temporary visa programs, the peculiar role of "Left-Right" coalitions in such policy advocacy, and the roles played in these policy debates by higher education and scientific and professional associations.

Confusion and Garbled Reporting about the Users and Actual Requirements for the H-1B Visa

Recent controversies about "shortages" have often focused on the H-1B temporary visa for "specialty workers." Chapter 2 includes a brief discussion of the origins of this visa, but the subject is one of such great confusion and misinformation that it warrants some clarification here.

Table 4.1. Employers with the Most New H-1B Visa Approvals, Fiscal
Years 2012 and 2011

Company	FY2012	FY2011
1. Cognizant	9281	5095
2. Tata	7469	1659
3. Infosys	5600	3360
4. Wipro	4304	2803
5. Accenture	4037	1304
6. HCL America	2070	930
7. Mahindra Group (incl Satyam)	1963	404
8. IBM	1846	987
9. Larsen & Toubro	1832	1156
10. Deloitte	1668	798
11. Microsoft	1497	1384
12. Patni Americas	1260	164
13. Syntel	1161	363

Source: Patrick Thibodeau and Sharon Machlis, "The data shows: Top H-1B users are offshore
outsourcers: U.S. government's H-1B visa list shows accelerating demand from offshore
outsourcers," *Computerworld*, February 13, 2013. Available online at http://www.computer
world.com/s/article/print/9236732/The_data_shows_Top_H_1B_users_are_offshore
_outsourcers?taxonomyName=Gov%27t+Legislation%2FRegulation&taxonomyId=70.
   Some company divisions were combined, such as IBM Corp. and IBM India, Tata consulting and engineering groups, etc.

It is truly remarkable how routinely the characterizations of these visas
put forward in political and media discussions differ in fundamental ways
from the facts, both in legal and administrative terms. There are many examples of such disconnects between rhetoric and reality, of which we summarize only three here.

First, as noted in chapter 2, these visas are often described as limited to
"high-skill" technical workers, sometimes described as "the best and the
brightest" in the world. Yet the facts are that the visas require education only
up to the level of a U.S. bachelor's degree or equivalent experience.

Second, because expansion of H-1B visa numbers has been so energetically promoted by U.S. companies such as Microsoft and Intel, many seem
to believe that such companies are the primary users of such visas. Again
the facts are otherwise: according to analyses of official data undertaken by
Computerworld magazine, by far the largest users of H-1B visas are companies that are "offshore outsourcers, many of which are based in India, or, if
U.S. based, have most employees located overseas."[1] (See table 4.1 for FY2012
and FY2011 data.)

The company in table 4.1 with the largest number of approved H-1B
visas in both years, Cognizant Technology Solutions, is illustrative of this

type of firm. Cognizant was established in 1994 and is now India's largest IT services company, having overtaken the previous leader Infosys in 2012 when Cognizant reported revenues of $7.35 billion and net income of $1.05 billion. Cognizant was created and incubated as a joint venture between another Indian IT services firm, Satyam (which later achieved notoriety for massive accounting fraud), and Dun and Bradstreet, a prominent U.S. company headquartered in northern New Jersey and a leading provider of business information.

Two years later Cognizant was spun off as an independent firm, with headquarters located in Teaneck, New Jersey. Cognizant is officially a U.S.-incorporated firm, included in the Fortune 500 ranking of the 500 largest American firms, and is listed on the NASDAQ. Moreover, nearly 80 percent of its revenues come from North America. However the large majority (published estimates vary from two-thirds to three-quarters) of Cognizant's 162,000 employees are located in its eleven IT service centers in India.[2] In addition, most of the minority of its employees who do work in the United States are Indian nationals on temporary visas such as the H-1B and L-1 (the latter is a temporary visa for intracompany transfers by companies with offices both in the United States and abroad). Indeed, Cognizant's legal filings with the U.S. Securities and Exchange Commission describe its business model as based heavily upon the use of H-1B and other temporary visas to bring its employees from India to the United States, where they facilitate offshore outsourcing of IT services to the firm's employees in India and provide other services:[3]

> Our future success will depend on our ability to attract and retain employees with technical and project management skills from developing countries, especially India. The vast majority of our IT professionals in the United States and in Europe are Indian nationals. The ability of Indian nationals to work in the United States and Europe depends on their ability and our ability to obtain the necessary visas and work permits ... There is a limit on the number of new H-1B petitions that United States Citizenship and Immigration Services, or CIS, ... may approve in any federal fiscal year, and in years in which this limit is reached, we may be unable to obtain H-1B visas necessary to bring foreign employees to the United States.

While most of the largest corporate users of H-1B visas listed in table 4.2 are firms of this type, the Computerworld article notes that

> [n]ot all of the major H-1B users are India-based ... Microsoft ranked 11th and has largely been the public face of those supporting a U.S. H-1B cap increase. IBM is also a major visa user but its numbers

also include the company's India-based operation. Global firms Accenture and Deloitte use the visa for IT services operations.[4]

A leading scholar of offshore outsourcing to India commented on these data as follows:

> There are two reasons these firms hire H-1Bs instead of Americans: 1) an H-1B worker can legally be paid less than a U.S. worker in the same occupation and locality; and 2) the H-1B worker learns the job and then rotates back to the home country and takes the work with him.... That's why the H-1B was dubbed the "Outsourcing Visa" by the former Commerce Minister of India, Kamal Nath ... Rather than keeping jobs from leaving our shores, the H-1B does the opposite, by facilitating offshoring and providing employers with cheap, temporary labor—while reducing job opportunities for American high-tech workers in the process.[5]

Third and finally, many express surprise when they learn that the H-1B visa, from the time of its creation in 1990 to the present, has never required any attempt by an employer to hire a domestic worker. Misunderstanding about this appears to be pervasive, due in part to the surprising frequency with which newspaper and magazine stories incorrectly state the very opposite of the facts—namely, that an H-1B visa can be issued only if the would-be employer can demonstrate a credible but unsuccessful effort to hire a domestic worker. This is simply garbled reporting: such a requirement has never been part of the H-1B visa process.

While we can never know why so many journalists have gotten this important point so wrong for so long, it may be due in part to credulous acceptance of statements to this effect by industry spokespeople or by other journalists. Journalists who are not experts in this area may be forgiven for misunderstanding U.S. immigration law, which is second in complexity only to the tax code. In particular, they may be confusing the H-1B (which is a "non-immigrant" or "temporary" visa) with permanent visas for employment such as the EB-2 or EB-3 visas[6] (often known as "green cards," which are "immigrant" visas providing legal permanent residence). Applications for most permanent immigrant visas for employment do require would-be employers to test the domestic labor market through a process known as "labor certification"—but there is no such requirement for the "temporary" H-1B visa. Even the terms used are confusing: the "labor certification" (LC) required for the EB-2 and EB-3 permanent visas for employment is rather easy to confuse with the "labor condition application" (LCA) needed for the H-1B temporary visa, but this LCA does *not* require the employer to make any effort to hire a U.S. worker.

Employer-supported Studies Claiming STEM Workforce Shortages

Over the past fifteen years or so there have been numerous studies, reports, testimonies, and other public pronouncements pointing to alleged STEM workforce shortages. Most of these have been sponsored or led by individuals or organizations in the corporate sector. This section considers the origins, contents, and effects of the most important and influential of these materials, some of which offer informative case studies of the use (and sometimes misuse) of data and projections about the U.S. science and engineering workforce.

In 1997, an employer association known as the *Information Technology Association of America (ITAA)* commissioned the first[7] of what was to become a multiyear series of reports raising concerns about what it called a large and rapidly increasing gap between demand and supply in the IT workforce. The 1997 report carried the nicely framed title of "Help Wanted: The IT Workforce at the Dawn of a New Century." It appeared during the early stages of a powerful boom of rapid growth in information technology (IT), a boom that was to prove unsustainable only four years later with a traumatic bust that began in 2001.

This 1997 report has been described as a "ground-breaking study" by ITAA's former president, and indeed it had significant policy impact. Oddly enough the report's cover and title page carried no indication of its authorship. A careful reading reveals that it was prepared under contract by Mr. Stuart Anderson, who at that time was director of Trade and Immigration Studies at the Cato Institute, a right-libertarian think tank in Washington. Anderson's own biography[8] indicates that he earned both a bachelor's and a master's degree in political science, but provides no indication of expertise in labor market analysis. His career illustrates pathways followed by ideologically committed policy advocates on both immigration and the science and engineering workforce.[9]

Anderson's 1997 report for the ITAA was based upon a proprietary (i.e., nonpublic) survey questionnaire he developed and sent to a sample he selected of one thousand employers in the IT sector and one thousand employers outside of the IT sector. One of the central questions addressed to respondents was how many "job openings" or vacancies they had on their books. Anderson's key conclusion was that there were some 190,000 such job openings in IT, and that this gap between supply and demand was greatly retarding economic advances in this industry sector.

Many questions were raised about the methods used in this study. The response rates to Anderson's proprietary survey were very low indeed—embarrassingly low. Only 12–14 percent of the two sample groups responded—a total of 271 responses from the two thousand companies to which the survey was sent. There was little reason to think that this small percentage of respondents was representative of the employers sampled. For example, it is plausible that the surveys' key questions about "job openings" might have received much higher response rates from those employers who actually were experiencing difficulty in hiring at the remuneration rates they were offering, but lower response rates from those that were experiencing few hiring problems. Moreover, as designed the survey's question about job openings surprisingly related to the companies' *global* workforces, rather than those in the United States—it is almost impossible to know how these data should be interpreted. Critics also noted that the study's estimate of 190,000 job openings was based upon a "grossing up" from what the 12–14 percent that responded had indicated were "job opening" numbers equal to about 10 percent of their companies' authorized workforces, but this "grossing up" calculation was based upon ITAA's own nonstandard definition of the size of the IT workforce.

The Anderson/ITAA study was also strongly criticized by experts on the IT workforce for failing to reduce its 10 percent estimate of "job openings" by the normal level of "frictional" job openings in the industry—that is, the normal percentage of positions not currently filled due to recent promotions, retirements, and departures. In the IT sector the percentage of all positions that are normally "open" for such reasons is usually reported to be about 5 percent of the total, and sometimes higher.

However serious may have been the weaknesses of the Anderson/ITAA study's data and methods, they did not seem to limit its effectiveness in the public policy sphere. Its claim of a large gap between supply and demand was widely reported in the press as a fact, and described as yet another indicator of the United States falling behind its economic competitors. The ITAA actively circulated the report to key members of Congress, and its worrisome conclusions were frequently cited by the press and in congressional testimony by spokespersons for the IT industry. Moreover, in the absence of any other quantitative evidence on the IT workforce, even the findings of a questionable study such as *Help Wanted* could not be challenged with more credible evidence. In an oft-cited slogan among Washington lobbyists, "in Washington politics, *any* number will always beat *no* number." ITAA's then-president made a similar point in different words: "our study may not be that good, but at least we have a study."[10]

In fall 1997, the Office of Technology Policy of the U.S. Department of Commerce published a report on the same topic, with the declarative title "America's New Deficit: The Shortage of Information Technology Workers."[11] Its findings were based in part upon the same 1997 ITAA report, along with the Bureau of Labor Statistics' ten-year projections of occupational employment published in 1995. The conclusions of this official report were far more cautious than those of the ITAA's *Help Wanted* report, but the Commerce Department report did refer frequently to "the shortage of IT workers," to apparent upward pressure on salaries, and to declines in the numbers of students graduating with computer science degrees.

In 1998, the ITAA produced a second "Help Wanted" study as a follow-up to its 1997 report. The response rate for this second report was also quite low at 36 percent, though far higher than the first report prepared by Stuart Anderson. Based on its methodologies, the second report indicated that the number of "open positions" in IT had grown from 190,000 to 346,000 between 1997 and 1998.[12]

These three reports attracted a great deal of attention in the press and in Congress, leading several members of Congress to request an evaluation by the General Accounting Office (GAO).[13] In March 1998, the GAO published a harshly critical assessment of both the 1997 and 1998 ITAA reports and of the related Department of Commerce report. The GAO report concluded that the reports' "shortfall" and "shortage" estimates were doubtful given the studies' weak methodologies and very low response rates. It noted, for example, that:

> We consider a 14-percent response rate to be unacceptably low as a basis for any generalizations about the population being surveyed. In order to make sound generalizations, the effective response rate should usually be at least 75 percent for each variable measured—a goal used by many practitioners. Furthermore, ITAA's estimate of the number of unfilled IT jobs is based on reported vacancies, and adequate information about those vacancies is not provided, such as how long positions have been vacant, whether wages offered are sufficient to attract qualified applicants, and whether companies consider jobs filled by contractors as vacancies. These weaknesses tend to undermine the reliability of ITAA's survey findings.[14]

Undeterred by such strong critiques of its first and second reports, the ITAA published a third report in 2000.[15] This time it concluded that the number of "open positions" in IT had increased to more than 843,000.

The harsh criticism from the GAO notwithstanding, the findings of the ITAA reports proved to be useful in the successful lobbying campaign by the IT industry that year in support of large expansion of the H-1B visa, the temporary-visa program for foreign "specialty workers." With strong support from Senator Abraham, who by then was chair of the Senate immigration subcommittee and had appointed Stuart Anderson (the author of the 1997 ITAA report) as its staff director, the industry lobbying campaign convinced Congress to triple the number of such visas, from 65,000 to 195,000 per year, for a three-year period.

The Lobbying Activities of Employers and Their Associations

ITAA later merged with other IT industry organizations to become a key part of a larger employer association called *TechAmerica*,[16] which has continued to lobby actively in this domain. Meanwhile the National Association of Manufacturers provided a base for another industry coalition, *Compete America* (initially named American Business for Legal Immigration, or ABLI), that lobbies actively in support of expansion of the H-1B visa program. Members of *Compete America* include major IT companies and their trade associations (e.g., Microsoft, Intel, Hewlett-Packard, TechAmerica), along with the leading American business associations including the Business Roundtable (CEOs of mostly large companies), the U.S. Chamber of Commerce (mostly small companies), and two leading associations of U.S. universities.

Another example of an industry association actively lobbying on these matters is the *Information Technology Industry Council (ITI)*, which describes itself as "the premier advocacy and policy organization for the world's leading innovation companies." Its members include nearly fifty of the largest international IT companies, including Microsoft, Intel, Oracle, Cognizant, IBM, Lenovo, Fujitsu, and Sony.[17] ITI's senior vice president for government affairs is Robert Hoffman, a former Senate staff member who later served as co-chair and spokesperson for *Compete America* while he was a vice president of Oracle Corporation,[18] and then as the chief Washington lobbyist for Cognizant Technology Solutions, the offshore outsourcing firm described earlier in this chapter as the largest single user of H-1B visas.

Other industry associations such as *Compete America* and the *Business Software Alliance (BSA)* have also lobbied directly and have retained the services of professional lobbying firms. The sources of financial support for *Compete America* are not publicly available, though it is commonly believed in the industry that Microsoft and Intel have provided much of its financial support. With respect to the *Business Software Alliance*, there has been open criticism within the software industry that Microsoft dominates both its funding and its activities.[19]

Companies such as Microsoft, Intel, and Oracle have not only supported lobbying by employer associations, but have also been engaged in direct lobbying efforts that have been amply financed and highly effective. They have deployed their own corporate lobbying organizations, and also have devoted many millions of dollars to retain numerous influential lobbying firms in support of increasing the numbers of both domestic and international scientists and engineers in the U.S. workforce.

At the time of the 1998 legislative debate that led to a near-doubling of the number of H-1B visas for example, Intel's lobbyists were rumored to have played a central role, but at the time this was only rumor and hearsay.

Twelve years later, in an odd twist of fate, contemporaneous documentary evidence about Intel's lobbying role in 1998 became public as an accidental by-product of the White House emails that were released in conjunction with the 2010 Supreme Court nomination of Elena Kagan, who had been serving in the Clinton White House in 1998.[20] The emails demonstrate that the Clinton administration initially had insisted that any expansion of the H-1B visa program then being sought by IT employers would have to be accompanied by other changes to mitigate its negative impacts upon U.S. high-tech workers.[21]

A May 1, 1998 email addressed to Kagan and Bruce Reed (the Clinton White House's chief domestic policy advisor and director of the Domestic Policy Council) from Julie A. Fernandes, a special assistant to the president focused on civil rights, race relations, and immigration,[22] recounted efforts to require companies seeking H-1B visas for positions offering average or below salaries to first attempt to hire U.S. workers; this requirement would have been waived for a new category of "H-1C" visas for positions that required "very high skill levels," defined as paying more than $75,000 (in 1998 dollars, equivalent to $106,000 in 2013 dollars). Such a requirement, wrote Fernandes, would call "industry's bluff re: their shortage of really highly skilled and desirable workers." Fernandes described Intel's lobbyist as "the most stridently opposed to our reform ideas when we met with high-tech lobbyists when this whole thing first started (a couple of months ago)."[23] The Clinton White House eventually dropped this proposal, and the H-1B visa has continued to require no attempt to hire domestic workers.

The Organization and Scale of Recent Lobbying Activities

Active industry-financed lobbying of this type has continued to the present. It achieved interim success when the goals of tech industry lobbyists were included as key components of the "Comprehensive Immigration Reform" legislation (known variously as the Kennedy-McCain bill and the Hagel-Martinez bill) that was debated in Congress in 2006 and 2007.

Ultimately these bills failed to win legislative approval, but as we shall see the industry responded by expanding its lobbying and related political efforts. Through the work of an "open government" organization called the Sunlight Foundation,[24] we now have access to much improved (though still not fully adequate) quantitative data about the magnitudes of such lobbying expenditures. Sunlight has compiled a searchable database of official lobbyist reports, categorized by lobbying issue, by client, and by quarter. It makes this database readily available to the public via the internet and without charge. Among its primary data sources are the quarterly Lobbying Re-

its lobbyists were engaged in lobbying on immigration issues for Microsoft alone.[32] (For more details on this firm and its founder and chairman, see later discussion in "Lobbying Firm Activities on Claimed STEM Shortages.")

In view of the failure of the Kennedy-McCain and Hagel-Martinez bills in 2006 and 2007, it is worth considering whether such aggressive lobbying efforts might have proved politically effective. On this subject there is an interesting quantitative study by three economists at the International Monetary Fund, published in 2008. The study found "strong evidence that both pro- and anti-immigration interest groups play a statistically significant and economically relevant role in shaping migration across sectors." The paper is limited in that it addressed only the number of visas issued for legal permanent immigration (i.e., it did not address temporary migration or unauthorized migration), and only for the period from 2001 to 2005. Its measure of business activity in support of increased immigration was based on business lobbying expenditures targeted on immigration policy, utilizing disaggregated data developed by the Center for Responsive Politics. The key finding drawn from these data was that a 10 percent increase in the size of lobbying expenditures by business groups per native worker is associated with a 2.9 percent larger number of [legal permanent] visas per native worker.[33]

Direct Lobbying by Senior Corporate Leaders

The visa lobbying activities of firms such as Microsoft and Intel is by no means limited to the industry associations whose lobbying activities they support, their internal lobbying organizations, or their outside lobbying firms. For both Microsoft and Intel, engagement has involved their most senior leaderships.

In the Microsoft case, Bill Gates, the company's primary founder, former CEO, and now chairman, and one of the world's two or three richest individuals, offered outspoken personal testimony of the subject to a Senate committee in 2007. A full transcript of Mr. Gates's testimony and of the questions senators posed to him is readily available online.[34] Gates urged the Senate to improve U.S. secondary education *and* to greatly increase admissions of foreign scientists and engineers under the temporary H-1B visa program:

> Now we a face a critical shortage of scientific talent. And there is only one way to solve that crisis today: Open our doors to highly talented scientists and engineers who want to live, work, and pay taxes here.... The fact is that the terrible shortfall in the visa supply

for highly skilled scientists and engineers stems from visa policies that have not been updated in more than 15 years.

When asked by a supportive senator how many temporary H-1B visas he thought should be provided, he replied:

> Well, my basic view is that an infinite number of people coming, who are taking jobs that pay over $100,000 a year, they're going to pay taxes, we create lots of other jobs around those people, my basic view is that the country should welcome as many of those people as we can get ...
>
> So, even though it may not be realistic, I don't think there should be any limit.

Mr. Gates's testimony was received warmly by the senators who attended this hearing, and indeed the Committee accorded him extended time to make his case. However, his comments evoked harsh responses from many scientists and engineers. The content and tone of such criticisms of Gates are illustrated, for example, in a blog published in a leading IT magazine, *ComputerWorld*, which described the exchange between Gates and his critics as "Round 4 of the H-1B war".[35]

Round 4: H-1B WAR—IEEE-USA vs. Bill Gates!

By Dino Perrotti
ComputerWorld
Created Mar 8 2007—12:55am

Bill Gates draws first-blood while IEEE raises an army of engineers in the pivotal fourth round of the H-1B battle of the Immigration Reform bill war.

In his opening move, the world's richest man strikes hard at American engineering careers by testifying before congress in favor of unlimited H-1B visas. Bill's prestidigitatory presentation sought to somehow convince lawmakers that the best way to encourage American students to enter engineering is to import more engineers from other countries.

First, he says, we must "place a major emphasis on encouraging careers in math and science." Then he suggests opening the flood gates to foreign engineers by eliminating the cap completely, "So even though it may not be realistic, I don't think there should be any limit." Sadly, there were no challenges to this point by the committee. Not one Congressman or Senator suggested that increasing

the cap might further discourage students from entering the engineering field.

Despite studies which show there is no shortage of engineers,
Bill Gates insists that the tech industry needs the cap raised.

Microsoft's chairman is perhaps the most visible corporate advocate of
such measures, but by no means the only one. Another prominent example
is Craig Barrett, a longtime leader of Intel Corporation who rose to become
its CEO and later its chairman. Like Gates, Barrett has been a strong advocate for both improving U.S. STEM education *and* increasing the numbers
of visas for foreign scientists and engineers. He was a member of the National Research Council committee that produced the *Gathering Storm* report
(see chapter 2), and also a member of the Council on Competitiveness. Dr.
Barrett's thoughts are available online in a fascinating debate on the website
TechCrunch[36] with Vivek Wadhwa, a software developer and executive who
is currently a vice president of Singularity University,[37] a regular columnist
for *BusinessWeek.com*, and an occasional writer for the online edition of the
Washington Post.

Both Barrett and Wadhwa provide intelligent and mutually respectful
arguments in support of their quite different perspectives. Barrett argues
that there are real shortages in the U.S. STEM workforce that encourage U.S.
employers to move such activities offshore, and that in response the United
States needs a major upgrade of its K-12 education to produce more high
school graduates who understand and appreciate STEM, more undergraduate majors in STEM fields, and more STEM graduate students. Wadhwa
replies that while he supports efforts to improve U.S. STEM education, there
are not really shortages in the U.S. STEM workforce, that the offshoring decisions by corporations are not due to U.S. educational or workforce deficiencies but instead to the favorable economics of moving such activities to
regions with lower-cost workers and more rapid economic growth, and that
the economic attractiveness of careers in these fields have been declining for
talented U.S. students who have more remunerative career options available
to them. In addition, the TechCrunch website provides some valuable comments and perspectives from nearly one hundred participants in the Tech-
Crunch blog.[38]

Lobbying Firm Activities on Claimed STEM Shortages

It is notoriously difficult to obtain hard information about the activities of
political lobbyists in Washington, almost all of which take place out of public
view. As we have seen from the incomplete but illuminating data provided by

the Sunlight Foundation, numerous lobbying firms are engaged in advocacy for changes in immigration law and policy. Most are retained by employers or employer associations, though some work for associations representing ethnic groups, professions, unions, etc. We cannot provide here a full mapping of the many firms so engaged. A few examples of those that currently are visibly active in such lobbying should suffice to illustrate their sources of support and the connections involved.

The Monument Policy Group LLC, whose founder and president is C. Stewart Verdery, provides lobbying and strategic counseling services to its clients—"well-known multi-national corporations, leading trade associations, local and state governments, and promising startups and non-profits," one of which is *Compete America*.[39] Verdery[40] is a former congressional staffer who from 2003 to 2005 served as assistant secretary for Border and Transportation Security Policy and Planning in the U.S. Department of Homeland Security (DHS), where he was responsible for "policies related to immigration, visas, and travel facilitation; cargo security and international trade; transportation security; and law enforcement."[41]

Corley Consulting is a newer and smaller lobbying firm founded in 2009 by Scott Corley, who earlier had been senior vice president at the Monument Policy Group and chairman of *Compete America*'s Advocacy Committee.[42] Before that Corley was director of Government Affairs for Microsoft Corporation, and previously had been director of Government Relations at the Information Technology Industry Council, described earlier. In 2012, Corley Consulting reportedly generated $630,000 in lobbying income from clients that included Microsoft, Intel, Immigration Voice, and Compete America.[43] Corley also now serves as the executive director of *Compete America*.

On the opposing side, a grassroots organization called NumbersUSA has retained three lobbying firms to further its legislative goals during the past five years: Timothy Rupli and Associates, led by a former staff member for the former House Majority Leader Tom DeLay (R-TX); Majority Group LLC (founded by former Idaho Congressman Walt Minnick); and the Mita Group. The Institute of Electrical and Electronics Engineers–USA, described later in this chapter, has retained the Morrison Public Affairs Group led by former Congressman Bruce Morrison (D-CT) to represent its opposition to proposed expansions of the H-1B temporary visa program.[44]

The Bizarre Case of Jack Abramoff and His "Team Abramoff"

One of the more surprising lobbying activities of leading IT companies was their engagement for many years of the now-infamous Washington lobbyist

Jack Abramoff, whom *Time* magazine called "The Man Who Bought Washington" and who described himself as "America's Most Notorious Lobbyist" in the subtitle of his own autobiographical book. Scandals about Abramoff's lobbying activities for the gambling operations of American Indian tribes had provoked a flood of media attention in 2005, and in 2006 he pleaded guilty to multiple federal charges of felony fraud, tax evasion, and conspiracy.

However there is surprisingly little mention in the mainstream media of Abramoff's apparently generously compensated representation of Microsoft Corporation. This lack of attention is doubly surprising because a quick web search on the topic would reveal numerous discussions of Abramoff-Microsoft links on many specialized websites focused on the high-tech industry.

Like lobbying activities in general, the details about Abramoff's lobbying for Microsoft remain murky. But there are a few matters for which there is well-documented evidence. One is that Abramoff was employed from 1994 to 2001[45] by Preston Gates Ellis & Rouvelas Meeds LLP. This Washington-based lobbying firm was a subsidiary of the Seattle law firm Preston Gates & Ellis LLP, a firm with offices in the United States, China, and Taiwan that was co-founded by Bill Gates's father and was very closely connected to Microsoft.[46] In this capacity Abramoff assembled a highly profitable lobbying group ("Team Abramoff") that was retained by Microsoft and by other clients.

Other close associates of Abramoff were also engaged separately as Microsoft lobbyists, including Grover Norquist, now president of the conservative grassroots lobby group Americans for Tax Reform and a major lobbyist in the Republican Party; Ralph Reed, until 1997 the executive director of the Christian Coalition and a prominent Republican Party activist; and William P. Jarrell, a former campaign manager and Deputy Chief of Staff to Rep. Tom DeLay, who later joined Abramoff at Preston Gates Ellis & Rouvelas Meeds LLP.[47] Both Norquist and Reed previously had worked under Abramoff in the 1980s when he was chair of the College Republican National Committee.

According to one source analyzing the lobbying filings made by the Preston Gates lobbying firm,[48] Microsoft paid Jack Abramoff $700,000 in 1998 to lobby for increased numbers of H-1B visas, another $560,000 in 1999, and an additional $480,000 in 2000. It needs to be acknowledged again here that it is notoriously difficult to obtain accurate information on the actual purposes for which lobbying funds are used. Real caution is in order in interpreting such numbers, but according to the firm's own regulatory filings there can be no doubt that Microsoft employed Abramoff and his colleagues as lobbyists on the H-1B issue.

As is now well known from his federal trial, Abramoff had very good access to the congressional leadership via his close ties with House Republican

Whip Tom DeLay. As mentioned previously, Abramoff's lobbying colleague William Jarrell had his own personal ties to DeLay. Abramoff's close associates Grover Norquist and Ralph Reed were two of the other most influential Washington lobbyists with special access to the Republican leadership, and both appeared to have benefited indirectly from considerable funds that Abramoff directed toward them and their organizations. In 1999, Reed was appointed directly as a lobbying consultant with Abramoff's lobbying firm.

In 1998, the legislative lobbying by IT employers and other lobbyists succeeded in increasing by 75 percent the original 65,000 numerical cap, to 115,000 per year. Sustained lobbying efforts over the next two years succeeded in October 2000 in tripling the original limit from 65,000 to 195,000 per year.

Just how much of this legislative outcome was due to Microsoft's financing of lobbying activities undertaken by Abramoff and his associates may forever fall into the category of the "unknowable," as it is nearly impossible to find clear documentary evidence of cause and effect from lobbying activities. Some well-informed observers believe the evidence of their impact is sufficient—their ready access and close ties to influential congressional leaders such as Tom DeLay are matters of record, as is the strong legislative support of the H-1B expansions by these same leaders. Other knowledgeable observers such as former congressman Bruce Morrison, the former chair of the House Immigration subcommittee, believe that energetic lobbying of Congressman DeLay and his associates was not really necessary because the latter were ideologically predisposed toward maximizing admission of temporary workers with few rights—as evidenced by their support for the extraordinary immigration policies of the U.S. Commonwealth of the Northern Marianas.[49,50]

For many years Jack Abramoff reigned as one of the most powerful and prosperous of Washington lobbyists, but in 2005 things began to fall apart. He was indicted in federal court—though not for his allegedly unlawful lobbying activities per se, for which enforcement is notoriously weak, but instead for a fraudulent business transaction in 2000 to finance the purchase of a cruise casino company called SunCruz. In a plea bargain he pleaded guilty to three felony charges of fraud, tax evasion, and conspiracy. In March 2006 he was sentenced to five years and ten months in federal prison.

Numerous associates of Abramoff's lobbying network were also indicted, and were forced to step down, to plead guilty, or were convicted of felonies. Both Tom DeLay (R-TX), by then the House Majority Leader, and Robert Ney (R-OH), chair of the Committee on House Administration, were forced to give up their leadership posts in the U.S. House of Representatives due to the scandal. Ney later pleaded guilty and was sentenced to thirty months in federal prison. Others such as Norquist, Reed, and Jarrell continue to operate as influential Washington lobbyists. Journalists at the *Washington Post*, who won a Pulitzer Prize for Investigative Reporting for their coverage of the Abramoff scandals, produced a useful timeline and list of principal actors in this political scandal, available online.[51] Abramoff presents his own version of the facts in his 2011 autobiographical book.

Other Associations and Organizations Involved in Lobbying

Nonprofit organizations too numerous to mention individually are actively involved in lobbying on immigration policies. These include, in addition to those already described as focused on skilled immigration, at least the following: Cato Institute; U.S. Chamber of Commerce; American Civil Liberties Union; Federation for American Immigration Reform; American Immigration Lawyers Association;[52] U.S. Conference of Catholic Bishops; Center for American Progress;[53] Americans for Tax Reform (led by Grover Norquist, the lobbying associate of Jack Abramoff); NumbersUSA, Immigration Works; and a large number of ethnic, religious, industry-specific, and other groups at national, regional, and local levels. Some of these have been active participants in "Left-Right coalitions" united only by common interests in immigration policies, to which we now turn.

The Peculiar Role of Left-Right Lobbying Coalitions

One of the more unusual aspects of lobbying and activism on immigration policy is that it cannot be understood along the conventional ideological dimensions of left or right, liberal or conservative. Nor does the standard partisan divide between Democrats and Republicans explain the positions taken, although party politics have become more important in recent years than in the past. This reality is understandably confusing to many journalists and other observers, who often refer to "liberal" or "Democratic" or "conservative" or "Republican" positions on immigration debates, terms that have little meaning on these matters.

Instead, immigration lobbying has been dominated by interest groups on the left and the right that have good connections with both political parties. These interest groups often find themselves in unfamiliar coalitions on immigration policy, supporting many policies in common although in pursuit of very different goals. The result has been unusual "Left-Right coalitions" on immigration, what many have termed "strange bedfellows."

To avoid any misunderstanding, we note that the term "Left-Right" as used here is not intended to suggest that such coalitions include groups along the full range of ideological positions from the radical left to the radical right. To the contrary—only some subsets of opinion on both the left and the right tend to be involved in these coalitions. For example, coalition participants on the right include right-libertarians such as the Cato Institute, Americans for Tax Reform, Club for Growth, and the *Wall Street Journal* editorial writers. More mainstream conservative groups, along with populist groups including "social conservatives" and Tea Party activists, are notable by their absence from such coalitions and sometimes for their opposition to them. On the left, coalition participants include left-libertarian groups such as the American Civil Liberties Union, some ethnic organizations, and some unions with large immigrant memberships, but include neither many groups normally associated with organized labor nor many traditional liberals.

What is distinctive about the Left-Right coalitions on immigration policy is that they combine groups and interests that disagree passionately on most other issues. These unusual political coalitions are real and active, but they did not self-assemble. To the contrary, they were assiduously organized by energetic advocacy groups and individuals.

One early example, and in some ways an archetype of Left-Right coalitions on immigration, is the National Immigration Forum,[54] founded in 1982 and originally called the National Forum on Immigration and Refugee Policy. Its founder, Rick Swartz, was a liberal/progressive immigration lawyer and advocate who previously had been active in promoting refugee status for migrants from Haiti. The primary source of funding for the Forum since its inception has been the Ford Foundation.

The Forum's co-founder and vice president, Phyllis Eisen, came from the generally liberal/progressive environmentalist group known as Zero Population Growth (which two decades later, in 2002, changed its name to Population Connection[55]). When she subsequently separated from the National Immigration Forum, she moved to the National Association of Manufacturers (NAM) as its lobbyist on immigration legislation, and later became NAM vice president in charge of its education, training, and research arm.

Swartz himself departed the Forum in 1990 to establish his own private consulting and lobbying organization known as Rick Swartz & Associates. Perhaps based upon his success in organizing the National Immigration Forum, Swartz subsequently has made something of a career specialty of assembling Left-Right coalitions.[56]

According to a laudatory article about Swartz in *Wired* magazine,[57] Swartz's first client was Richard Gilder, founder and chair of Gilder, Gagnon, Howe & Co., a Wall Street firm that has profited greatly from its aggressive trading of leveraged stocks and short-selling. Gilder was also an ownership partner of the Texas Rangers with George W. Bush. With his substantial wealth, Gilder has long been a major benefactor of conservative and right-libertarian think tanks. For example, he was the founder and board chair of the Club for Growth, a politically aggressive right-libertarian group that finances political challengers of Republican politicians who do not share its goals.[58] He also has been a Board member and large financial contributor to the Manhattan Institute.

Given their divergent political commitments, Gilder and Swartz were likely on different ideological planets on most topics. But they apparently were in synch on the desirability of expansive immigration policies, and Gilder financed Swartz to put together Left-Right coalitions that would support such immigration policies as well as free trade. According to a report[59] produced by CIP Americas, a think tank based in Mexico City that is critical of Swartz's activities from a left-wing perspective, during the early 1990s Swartz successfully assembled such a Left-Right coalition in support of the North American Free Trade Agreement. Later he coalesced a less successful Left-Right coalition to oppose the continuation of U.S. farm subsidies. In the mid-1990s he helped the National Immigration Forum that he had founded pull together a Left-Right coalition of large corporations, business associations, Silicon Valley firms, right- and left-libertarian groups, and ethnic and religious groups into an otherwise improbable alliance that mobilized opposition to proposed efforts to reform U.S. immigration laws and make them more enforceable.

It is interesting to note that both this CIP Americas article and the earlier article in *Wired* magazine reported that as Swartz was pulling together this Left-Right coalition on immigration, he was also serving as a consultant for Microsoft on skilled-labor immigration policy.[60] Both articles also noted that a core member of the unusual coalition was the same Phyllis Eisen, Swartz's former partner at the National Immigration Forum and later a vice president at the National Association of Manufacturers.

The *Wired* article approvingly recounts the legislative successes of the 1995 Left-Right coalition assembled by Swartz. It quotes him as describing himself as "the broker, the head banger, the guy who gets in and mixes it up." The coalition he assembled on immigration consisted of a disparate assemblage of advocacy groups, many of whom were also members or supporters of the National Immigration Forum that he had created in 1982. This group met every two weeks to discuss how to work together to achieve what may have been its only goal in common: expansion in the numbers in one or all categories of U.S. temporary or permanent visas. The *Wired* article catalogs some of the organizations involved in this Left-Right coalition, which it described as "by any standard, a crew of staggering ideological eclecticism":

• American Civil Liberties Union (the nation's leading left-libertarian activist group)

- Center for Equal Opportunity (right-libertarian think tank led by Linda Chavez)

- National Conference of La Raza (a leading Latino activist group)

- Organization of Chinese Americans (ethnic activist group, later renamed OCA)

- U.S. Conference of Catholic Bishops (the hierarchy of the U.S. Catholic Church)

- American Jewish Committee (a leading ethnic Jewish think tank and advocacy organization)

- National Association of Manufacturers (sponsoring organization for Compete America, the coalition of IT employers lobbying for expansion of the H-1B visa)

- National Federation of Independent Businesses (a large lobbying organization for small firms)

- Americans for Tax Reform (right-libertarian lobbying organization led by Grover Norquist)

- Cato Institute (right-libertarian think tank). Apparently the Cato Institute was represented in this coalition by Stephen Moore, who later left Cato to found the Club for Growth, a lobbying group with an affiliated political action committee that was founded and funded by the same Richard Gilder who had financed Rick Swartz's new consulting firm. Moore subsequently has moved on to become a member of the right-libertarian editorial board of the *Wall Street Journal*.

While indeed it might be "a crew of staggering ideological eclecticism," this combination of a coalition of large IT employers willing and able to finance lobbying and campaign contributions to supportive politicians, activist organizations on both the left and right, and ethno-religious lobbying organizations appears to have been a formidable lobbying force. In the mid-1990s it helped to block legislative changes in the family immigration system recommended unanimously by the U.S. Commission on Immigration Reform chaired by the late Rep. Barbara L. Jordan, by all standards a liberal and civil rights icon.[61] Similar Left-Right coalitions have been active in blocking numerous reforms in the U.S. immigration system recommended by many other centrist groups and bipartisan commissions and committees.

With respect to political debates about the science and engineering workforce that is the focus of this book, the Left-Right coalition created by Swartz provided at least the appearance of broad-based political and interest-group support for the policies promoted by their high-tech corporate supporters. It remains unknowable to what extent its lobbying activities turned out to be decisive.[62] Still it seems plausible that their efforts contributed—in some admittedly hard-to-quantify way—to the resounding success of the efforts led by IT employers to expand the numbers of temporary workers admitted under the H-1B visa system.

The Roles U.S. Higher Education Played

U.S. institutions of higher education have not been among the leaders of active lobbying on immigration matters, though some have chosen to become supporting members of organizations lobbying for increasing immigration. As noted earlier, several leading organizations of higher education are among the otherwise mostly business members of *Compete America*, the organization at the National Association of Manufacturers that lobbies for expansion of the H-1B visa program. These include the Association of American Universities (AAU—the association of the sixty leading U.S. research universities plus two in Canada) and the Association of Public and Land-Grant Universities (APLU). Still, the role of these higher education organizations appears to be mostly in the form of lending political support rather than leadership on immigration advocacy.

However, one higher education association has been among the most active lobbying organizations about some immigration matters. Its activities during the late 1990s were to become embarrassingly tangled in events related to the September 2001 Al Qaeda attacks on New York and Washington. This association is called, somewhat awkwardly, "NAFSA: Association of International Educators."[63] It is a membership organization engaging some eight thousand employees of U.S. educational institutions who provide advice and assistance to international students and scholars.

In 1996, legislation was passed to reform enforcement and oversight of the high-volume foreign student visa system, administered by the Department of State and Department of Justice. The goal of this legislation was to strengthen previously lax enforcement of the admission and temporary residence of hundreds of thousands of foreign students each year. NAFSA energetically opposed this legislation, and after its passage in 1996 decided to mobilize its lobbying efforts to block its implementation. As we shall see, it succeeded in doing so for more than five years—until 2001. After the September 2001 attacks on the World Trade Center and the Pentagon, it emerged that several of the nineteen Al Qaeda hijackers had taken advantage of the continuing lax enforcement and oversight of student visas to stay in the United States in pursuit of their suicide mission.

There is a long and convoluted story here, chronicled in a fascinating investigative journalism story by Nicholas Confessore (now a *New York Times* journalist) published in the *Washington Monthly* in 2002. The title and subtitle of this story summarize its findings: "Borderline Insanity: President Bush wants the INS to stop granting visas to terrorists. The biggest obstacle? His own administration."[64]

Concerns about use of student visas for entry to the United States by terrorists had increased after the first (and nearly successful) effort in 1993 to collapse the World Trade Center in New York City using a truck bomb parked in an underground garage near the building foundations. The truck driver was a Palestinian who had entered the United States on a student visa in 1989, briefly attended Wichita State University, and then dropped out to work in Texas. This change in his student status should have led to cancellation of his student visa and his departure from the United States, but it was not reported to the Immigration and Naturalization Service (INS) of the Department of Justice due to the long-standing lax monitoring of student visas.

In the following year a task force in the Department of Justice, led by an INS official named Maurice Berez, examined the foreign-student visa program in detail. In its final report in December 1995, the task force found that international students were "not subject to continuing scrutiny, tracking or monitoring when they depart, drop out, transfer, interrupt their education, violate status, or otherwise violate the law."[65] It recommended a new foreign-student tracking system based on computer records rather than paper forms, which would allow applicants for student visas to be routinely and quickly checked against law enforcement databases on terrorists and money-laundering. Moreover, once a student visa was issued this automated system would enable schools and universities to efficiently provide reports via the Internet to confirm that the student had indeed enrolled and continued to be a student—thereby reducing the ability of student visa-holders such as the 1993 truck bomber to violate their visas by dropping out and disappearing into the workforce or, as in his case, into a terrorist underground.

The realities of the near-miss in the 1993 World Trade Center attack seemed to have mobilized unusually rapid action by the U.S. government, and in the 1996 legislation the new Internet-based student visa system proposed by Berez's task force was mandated by Congress and christened with a new Washington acronym, CIPRIS (Coordinated Interagency Partnership Regulating International Students).

According to Confessore's report, the CIPRIS system had substantial support from prominent members of NAFSA, many of whom had assisted the INS in developing and pilot-testing the new Internet-based system. They saw it as offering real improvements of the existing paper-based system, which was not only ineffectual but also administratively burdensome. Confessore quotes Catheryn Cotten, a prominent NAFSA member and director of the foreign-student office at Duke University, as follows about the CIPRIS system:

> We love it ... , [Berez] understood the big picture. But he also knew to talk to people on the ground who did this every day. He was really creating a package that would serve a lot of purposes, including ours, in the easiest and most efficient way.

While the CIPRIS system evidently was supported by many NAFSA members, the organization's leadership was fiercely opposed. Under their instructions, NAFSA's lobbyists first used their influence with senior officials at the INS to cause Berez to be removed from the CIPRIS project and reassigned, and then sought to delay implementation of the system mandated by the 1996 law. They decided to do so by challenging the $95 fee that the legislation called for to finance the new system as "an unreasonable barrier to foreign students" and as an unacceptable federal mandate upon universities to collect it.

In these efforts they found a strong ideological supporter in Stuart Anderson, the former Cato Institute staffer who had written the 1997 ITAA report "Help Wanted," discussed earlier in this chapter. By then Anderson had become immigration policy director for the chair of the Senate immigration subcommittee, Senator Spencer Abraham, a Republican freshman from Michigan with close connections with the Cato Institute. Abraham and Anderson organized a group of twenty-one senators to ask INS commissioner Doris Meissner to delay further implementation of the CIPRIS system until NAFSA's objections to the fee system could be resolved. Meissner agreed and put the project into suspension.

Confessore reports that while implementation of the CIPRIS system was thus suspended, its opponents within the government took steps to weaken the system by eliminating its ability to cross-check information with the other databases on terrorists and money-laundering, by sharply reducing the amount of information to be collected, and by eliminating the CIPRIS system's counterfeit-resistant visa cards. Thus modified, the new and reduced version of CIPRIS, still in suspension, was renamed SEVIS (Student and Exchange Visitor Information System).

It was all to go badly wrong. While CIPRIS and its reduced-form version SEVIS were still languishing without implementation in 2000 and 2001, three of the nineteen suicide attackers (including three of the four pilots) who succeeded in conducting the 9/11 attacks had obtained student visas that authorized them to stay in the United States—though for reasons other than education.[66] Their fellow 9/11 conspirators took advantage of other poorly monitored temporary visas such as those for tourism and business to enter and stay in the United States.

Within only a few days after the September 11 attacks, evidence of this use of the weakly enforced student visa system by these suicide terrorists appeared in press coverage. NAFSA found itself under strong pressure from more influential Washington-based higher-education associations that had previously supported NAFSA's lobbying positions. Within days NAFSA announced that it would no longer oppose implementation of the SEVIS system, and with NAFSA's opposition withdrawn, the system was removed from suspension and implemented. It is impossible to know whether, had it been implemented three years earlier as planned, SEVIS's automated reporting mechanisms would have attracted the attention of the INS or security agencies to the use of student visas by three of the four Al-Qaeda pilots of the hijacked planes used in the 9/11 attacks.

Six months later one of the more bizarre and embarrassing sequelae to the September 2001 attacks on New York and Washington hit the press— the INS suddenly delivered formal letters of approval of the student visas for Mohamed Atta and Marwan Al-Shehhi, two of the four aircraft pilots so central to the success of the 9/11 suicide attacks. These visa approval letters were sent to the Florida flight school at which the two Al Qaeda operatives had long since completed the flight training they had needed to transform the hijacked planes into (human) guided missiles. In response to questions at a press conference shortly thereafter, President George W. Bush famously recounted that when he was informed about the two INS approval letters for Atta and Al-Shehhi's student visas, "I was stunned and not happy. I could barely get my coffee down." Attorney General John Ashcroft pointed to "professional incompetence." Even the INS Commissioner James Ziglar called it an "inexcusable blunder."[67]

One might well ask why the major Washington-based higher education organizations were supportive of NAFSA's successful efforts to block implementation of the new student visa system. International student/scholar advisors hardly are influential in policy matters on most campuses, and their national association NAFSA is similarly not very influential among higher education associations.

This is another area in which the answers are likely unknowable. One plausible explanation was that the leaders of other higher education organizations deferred to NAFSA as the organization with the greatest expertise on international student visas. Another plausible explanation reflects the views openly expressed by many in academe—that U.S. universities must expand international student enrollments in response to rising international competition from other countries' higher education systems, for reasons that include: generation of tuition revenues, especially in net revenue programs such as business degrees; filling of classes in graduate programs that are attracting insufficient domestic student interest; international graduate students in U.S. science programs as a source of research assistants for externally funded projects; extension of "diversity" goals to the international sphere; and U.S. institutions' responsibility to provide higher education to the world. Empirically, we know that most higher education organizations have applauded the expansion of international student populations on their member campuses, and have expressed alarm when such flows have turned downward. Hence the support for NAFSA's lobbying efforts could be said to be at least consistent with the perspectives of many influential organizations in U.S. higher education.

There is, finally, yet another question to contemplate for which the answer is likely unknowable: at what point will the federal government become effective in regulating fraud and abuse in the visa categories related to international students? In spite of the deeply embarrassing failures just described, it appears that the office that now is responsible for the SEVIS and related visa programs—the Department of Homeland Security's Immigration and Customs Enforcement (ICE)—has replicated the ineffectual management of student visas by the INS. According to a 2012 report to the Senate Subcommittee on Immigration by the Government Accountability Office (GAO), ICE "has not developed a process to identify and analyze program risks since assuming responsibility for the Student and Exchange Visitor Program (SEVP) in 2003." ICE has allowed the number of institutions licensed to participate in this visa program to expand to a number that is very difficult to monitor—more than 10,000—including more than 1,000 that are not even accredited. The GAO report found that "ICE officials do not consistently verify certain evidence initially submitted by schools in lieu of accreditation. In addition, ICE does not maintain records to document SEVP-certified schools' ongoing compliance."[68]

Research Universities and the H-1B Visa

As noted earlier, immigration issues other than those related to student visas have not been a priority issue for those who advocate for research universities, who have focused their efforts primarily on obtaining increased federal funding for university-based research, education, and student loan programs. In 2000, however, lobbyists for the IT industry actively sought support from research university lobbyists for the legislative efforts then under way that ultimately succeeded in tripling the number of H-1B visas, from 65,000 to 195,000 per year. Initially the lobbyists for research universities demurred, noting that there would be little advantage for their university clients since most of the additional visas would be used by companies because of the way these visas were allocated by the government.

Evidently the IT lobbyists believed that political support from universities would be helpful in passing the H-1B expansion legislation they were promoting. They responded by offering to modify the legislation to completely "uncap" the number of H-1B visas available to universities and non-profit research institutions; universities would no longer have to compete with companies in obtaining H-1B visas within the numerical limit. This was an offer that the university lobbyists could not refuse, and they signed on to support the revised legislation proposed by the industry, which was passed in 2000.[69]

Later an additional tranche of 20,000 H-1B visas per year was made available to foreign students completing master's or higher graduate degrees in U.S. universities, a further expansion also supported by university lobbyists.

The Weak Roles of Professional Societies and Unions Representing Scientists and Engineers

In most mixed economies including the United States, organizations of employees routinely negotiate directly with employers over issues such as pay, benefits, pension rights, working conditions, seniority, layoffs, etc. Employee organizations also press their interests in the political and policy domains, seeking legislative and/or regulatory advantage by providing candidates with campaign finance, political advertising, endorsements, and substantial volumes of volunteer labor (e.g., the campaign telephone banks often staffed by unions).

Such activities are less common in the professions, although they are actively pursued by some of the largest professional associations, e.g. the two large and active national unions representing public school teachers.[70] Non-union professional organizations representing medical doctors (American Medical Association) and lawyers (American Bar Association) also engage in significant efforts to represent the economic interests of their memberships.

However, such organized representation of the interests of scientists and engineers is far less evident. This is not due to a paucity of large professional organizations or societies. To the contrary: science and engineering fields support many national disciplinary and professional organizations, some of which have large memberships. In engineering, for example, there are more than twenty-five professional societies, institutes, and associations in the United States, mostly configured along disciplinary or industry lines. The largest is the Institute of Electrical and Electronics Engineers (IEEE), which reports about 400,000 members, more than 1,200 staff members, and annual expenditures that exceed $300 million.[71]

However, most engineering societies including IEEE consider their roles as primarily technical. IEEE publishes some 130 journals and related technical publications each year, and sponsors 800 conferences on technical engineering issues. It has developed and maintains hundreds of international engineering standards for fields as diverse as information technology, power and energy, telecommunications, networks, nuclear engineering, and software engineering.

IEEE is a global professional and technical society, and as such understandably has not been much involved in issues affecting the attractiveness of engineering careers in the United States. Its corporate headquarters are located in New Jersey, but its 400,000 members are in 160 countries. Its thirty-one-member board of directors reflects this by reserving ten directorships for representatives from its ten regions around the world. Hence it is not surprising that the IEEE as an organization is not known for its representation of electrical and electronic engineers in any particular country, including the United States.

However, over half of IEEE members are located in the United States—some 210,000 of the IEEE's global membership of 400,000.[72] In view of this membership concentration in one country, the IEEE has a U.S. subsidiary known as IEEE-USA with offices in Washington, DC, that provide membership services to its U.S.-based members. In the past, IEEE-USA has sought to represent the economic and career interests of U.S. engineers, for example by criticizing the effects of the H-1B visa program upon U.S.-based electrical and electronic engineers. However, such efforts led to conflicts with leaders of the parent IEEE organization, who sometimes have imposed restrictions on U.S. advocacy by IEEE-USA as contrary to the interests of the IEEE as a global organization.[73] Suffice it to say, therefore, that global engineering associations such as IEEE have not focused on domestic policy issues that may affect the economic and professional well-being of their members in any specific country, including the United States.

In the sciences, the number of professional societies in the United States is even larger than in engineering, but these associations tend to be far smaller and are often highly specialized by discipline and subdiscipline. The Council of Scientific Society Presidents lists more than fifty national scientific societies among its members,[74] and with only modest overlap the Federation of American Societies for Experimental Biology (FASEB) has some twenty-three member societies in the biological and biomedical sciences.[75]

Among this multitude of scientific and engineering societies there is a broad consensus as to the importance of science, engineering, and mathematics, often reflected in support for increased government funding for research and education in these fields. Many of these societies or federations support active "public affairs" functions that emphasize the importance of science and engineering to the public good. On such matters there is unity in purpose if not in specific actions.

However, unlike other professions such as medicine, law, or K-12 teaching, the many fields of science and engineering are conspicuously lacking in any national umbrella organization that pays regular attention to the careers and other interests of their members. There is no equivalent in the sciences or engineering to the American Bar Association or the American Medical Association, not to mention the large and even more politically active unions that have been formed by public school teachers.

In the sciences, the closest to such an umbrella organization is the American Association for the Advancement of Science (AAAS), the publisher of *Science* magazine. AAAS has more than 100,000 individual members, along with loose affiliations with many scientific disciplinary societies. In engineering, the closest to a national umbrella organization is the American Association of Engineering Societies (AAES). This organization actually has no individual members; instead it is a loose coalition of some twenty-seven independent engineering societies. Neither of these umbrella organizations has been heavily involved in debates related to alleged "shortages" of scientists and engineers, nor have they taken strong stands on public policy proposals that might affect the career prospects of scientists and engineers as broad occupational groups.

Under these circumstances—strong and well-financed lobbying by employers of scientists and engineers and their employer organizations, combined with scattered, hesitant, and sometimes conflicted engagement by fragmented professional organizations of scientists and engineers—there is really no contest. The outcomes of public policy debates unsurprisingly have been largely consistent with the wishes of the employer groups.

Summary and Conclusions

Over the past two decades, lobbying and public relations efforts to convince U.S. political elites that the country faces damaging and widespread shortages in its critical science and engineering workforce can only be described as stunning successes. It is conventional now to hear seemingly sincere pronouncements about the dangers of such shortages from politicians of all ideological persuasions and from much of the mass media. This apparently broad consensus prevails notwithstanding almost universal inability by objective labor market analysts to find any convincing empirical evidence to confirm the existence of such generalized shortages.

These successful efforts to mold political and public opinion have been led and largely financed by employers, employer associations, and their lobbyists. Their most notable successes have been the creation and then great expansion of temporary visa programs such as the H-1B, the requirements and beneficiaries of which have been widely misconstrued in the press and policy discussions. Supporting roles have also been played by immigration lawyers, organizers of unusual Left-Right coalitions, and some higher education leaders and lobbyists.

Advocates of these shortage claims have had a nearly open field in politics and the media. Some groups have challenged the shortage claims and pointed to the absence of credible evidence, but they are far less well organized, funded, and politically connected. There are numerous large associations of scientists and engineers that include many members who question such shortage claims, but these professional associations often abjure political advocacy as a matter of principle or are constrained by internal conflicts.

CHAPTER 5

What Is the Market Really Like?
Supply, Demand, Shortage, Surplus—and Disequilibria

> History is strewn thick with evidence that a truth is not hard to
> kill, but a lie, well told, is immortal.
>
> —Mark Twain

Much of the debate about the adequacy of the U.S. science and engineering
workforce is dominated by contesting claims of "shortages" or "surpluses."
What do these terms actually mean? How are they used by advocates and by
others in politics and the press? How would an objective observer go about
assessing such dueling claims?

Such questions lead immediately to a surprising reality: even some of the
most basic questions about labor markets in science and engineering have
multiple, and quite different, answers. Among the questions producing conflicting answers are the following, many of which one might think could be
answered with clarity:

- What is meant by a "shortage" or "surplus" of scientists and
engineers?

- How large is the U.S. science and engineering workforce?

- What is meant by "high-tech" industries and the "high-tech"
workforce?

- Are there general "shortages" or "surpluses" in demand relative
to supply?

- Are there particular fields, industries, or regions for which supply and demand are not in balance?

- How should we understand the significance of trends and rates in PhD production?

- How does the number of recent PhDs compare with PhD career opportunities in the United States?

- How many scientists and engineers are working in the United States on temporary visas?

- How many science and engineering "postdocs" are there in the United States?

What Is Meant by a "Shortage of Scientists and Engineers"? (or a "Surplus"?)

Part of the deep confusion and dissension about claims of "shortages" (and sometimes "surpluses") of scientists and engineers is straightforwardly semantic. The same English words of "labor shortage" and "labor surplus"—but with very different meanings—are used by economists who study labor markets on the one hand, and by employers, politicians, policy advocates, and journalists on the other.

Most economists would agree with the definition of "labor shortage" in a market economy used by the Bureau of Labor Statistics (BLS): "when the demand for workers for a particular occupation is greater than the supply of workers who are qualified, available, and *willing to do that job*."[1] In the same vein, a recent book by Burt Barnow, Jaclyn Schede, and John W. Trutko concisely summarizes the concepts of both "surplus" and "shortage" as follows:

> [If] "the quantity of labor offered exceeds the quantity that firms wish to purchase, there is a surplus, and if the quantity of labor desired by firms exceeds the amount workers offer *at the prevailing price*, there is a shortage.[2]

There is broad consensus among labor economists about these definitions. There is also consensus that labor markets in mixed economies will adjust to such circumstances via changes in both supply and demand.[3]

In public policy discussions, however, confusion often prevails because precisely the same English word "shortage" is used in a very different way. This occurs most commonly when there is a period of increasing labor market demand in a particular occupation. Employers who have been accustomed to easily hiring employees with certain qualifications at a particular wage level begin to find themselves facing more competition, and hence increased difficulty in filling their recruitment goals at the wages they are accustomed to offering. They describe this as a "labor shortage," by which they mean simply an insufficiency in the number of qualified candidates at those wage levels. Often they go on to argue that there is need for an increase in the supply of people who are willing to work under the terms they are offering, rather than a need for employers to adjust to the market through changes in remuneration, recruitment, and work organization.

Hence the primary definitional or conceptual difference between the perspectives of labor market experts and employers is that employers often seek more labor supply so they do not have to increase wages or make other market adjustments. Meanwhile, most labor economists would describe situations in which rising demand results in insufficient numbers of qualified candidates willing to work at the wage and working conditions being offered as reflecting a market disequilibrium requiring multiple adjustments.

A classic 1957 economics study on the subject, co-authored by the Nobelist George Stigler, describes expanding the supply of qualified employees as only one such adjustment. Others include: enhanced recruitment efforts by employers; increases in the remuneration offered; scrutiny and appropriate modification of the qualifications that employers may have become accustomed to requiring (e.g., less detailed specifications of required experience or educational backgrounds); upskilling of their existing workforce via expanded training; and restructuring of the employers' workforce by, for example, reassigning some portion of the work to different categories of workers or expanded use of overtime.[4]

In short, the same English phrase, "labor shortage," is used to convey very different meanings—an obvious recipe for confusion. These different meanings in turn suggest very different responses. Employers often focus primarily on the *supply* side, urging that someone (the government? the education system?) provide more candidates by increasing the number of graduates and/or visas for foreign hires. Labor economists tend to emphasize a need to address *both the supply and demand sides*. Unsurprisingly the outcome of public debate about a seemingly simple English word—"shortage"—with such multiple meanings and implications is dissonance and confusion.

Other Meanings of "Labor Shortage"

There are also a few other usages of the term "labor shortages" that deserve brief mention here. In the "Social Demand Model," a shortage of scientists and engineers would exist if people *believe* that the economy (or some other aspect of society) would be better if there were more of them, even if the labor market is accurately reflecting market demand. In their classic 1959 paper on "shortages" of scientists and engineers then being claimed (see chapter 2), Kenneth Arrow (also a Nobelist) and William Capron summarize this perspective as follows:

> In particular, careful reading of such statements indicates that the speakers have in effect been saying: There are not as many engineers and scientists as this nation should have in order to do all the things that need doing such as maintaining our rapid rate of technological progress, raising our standard of living, keeping us militarily strong, etc. In other words, they are saying that (in the economic sense) demand for technically skilled manpower *ought* to be greater than it is—it is really a shortage of *demand* for scientists and engineers that concerns them.[5]

Another different usage of "shortage" comes from the obvious reality that there can be only a limited number of truly creative geniuses and leading-edge innovators. These are the *truly* best and brightest—those few individuals with unique capabilities, who by definition must be small in number and who predictably will be in high demand. For true geniuses and innovation leaders, "shortages" are inevitable and permanent.

Types of True Labor Shortages

Labor economists do acknowledge that true labor shortages of scientists and engineers can occur, and point to three distinct categories—static, dynamic, and structural.[6]

Static shortages and surpluses: This most widely used notion of "shortages" or "surpluses" of scientists and engineers would occur if, based on a snapshot at a point in time with current labor market conditions and prevailing remuneration rates, there are fewer (or more) qualified scientists and engineers than there is market demand for them.

Of course any such snapshot can be misleading, since over time such markets would be expected to adjust by changes in both demand and supply. If employers experience such a "shortage" in a particular field of science or engineering, the market would cause them to adjust by for example improving their remuneration offers, thereby encouraging additional entrants attracted by promising career prospects. If, on the other hand, scientists and engineers find themselves in labor markets with surpluses in supply relative to demand, market forces would lead some to adjust by departing these occupations for other more attractive ones, and the number of new entrants would decline due to unpromising career prospects.

Dynamic shortages and surpluses: Adjustments of the kinds described previously would be expected on the basis of standard economic theory, but, like nearly everything in economics, predictions based upon theory may not work out in practice. Consider what would happen if there was sudden growth ("boom") in demand for such personnel, accompanied by long delays in the "production" of additional qualified scientists and engineers—under such conditions the supply response of new entrants might not be as rapid as employers might hope and wages would be expected to increase.[7]

Similarly, sudden declines in demand (a "bust") would provoke departures by some scientists and engineers (or soon-to-be-qualified students) experiencing an over-supply situation, but their departures might be slower than the decline in demand in view of the large investments they had already made in preparing for such careers. This is especially important when there are *cyclical* shortages and surpluses, driven by upward and downward cycles in demand. In such cases, participants may remain in the field in hopes they can wait out a temporary downturn, an option that becomes more plausible if they have ready access to alternative sources of temporary remuneration. Here the availability of funded postdoctoral positions, however unattractive in economic or career terms, may have become an important market "reservoir" or "holding tank" for new PhD scientists experiencing poor initial prospects.

Structural shortages and surpluses: A third kind of shortage or surplus could arise if wages and other remuneration do not respond promptly to changing labor demand and supply conditions. Some employers may legitimately be unable to raise wages sufficiently to attract the labor supply they need because their domestic competitors are tapping low-wage workforces abroad or using temporary visas to import such workers. Under these conditions, increased domestic supply would not be stimulated and labor market competition would impel more employers to emulate the practices of their lower-wage competitors, including both offshore outsourcing and use of temporary foreign workers.

Another "structural" example would be employers (private or public) whose remuneration offers are limited by political or contract provisions rather than by markets. Examples of politically determined limits would be reimbursement rates set by Medicare or Medicaid. A different example would be remuneration rates negotiated under union contracts that prohibit differentially higher pay for secondary school teachers in disciplines with stronger demand than others, such as science and mathematics as compared with art or music.

The federal government may face its own special kind of structural shortage in specific areas of science and engineering during periods when there is robust demand for such expertise outside of government. Private employers may have the flexibility to increase remuneration to attract supply, while governmental employers may be constrained by official civil service pay grades and by other hiring requirements such as preferences for military veterans.

Finally, in a few specific industries related to national security that employ significant numbers of scientists and engineers, both government agencies and firms may find themselves facing special recruitment problems if substantial fractions of the potential workforce from which they might recruit are foreign nationals who cannot readily obtain the security clearances needed for such employment. Examples here would include recruitment into sensitive positions (e.g., nuclear physicists, mathematicians, aeronautical and electronic engineers, cybersecurity) of international students by Department of Defense agencies, intelligence agencies, and their "ecosystem" of private sector contractors. It is perhaps worth noting here that the large security breaches admitted by U.S. citizens holding such security clearances, including Bradley/Chelsea Manning and Edward Snowden, must raise real questions about the credibility of the security clearance process.

Efforts to Model Science and Engineering Personnel "Requirements": Over-interpretation

An adequate supply of scientists and engineers is important for the economic prosperity, educational quality, and national security of the United States and other advanced economies. Industries and firms that depend upon active research and development or upon products emerging from such activities are important contributors to economic prosperity, and could face difficult competitive positions if they were unable to recruit sufficient numbers of employees with the relevant expertise. Systems of higher education also require a sufficient supply of highly advanced personnel to serve as the faculty and research staff who educate undergraduate and graduate students and conduct research. The national security domain too has become increasingly dependent upon science and technology-intensive capabilities, and in this case there is a further requirement of a rough sufficiency among the citizen workforce able to obtain needed security clearances. Major insufficiencies in supply of scientists and engineers would pose significant challenges for all of these.

At the same time, it is both wasteful and counterproductive to pursue a policy that results in large excesses in the supply of such workforce capabilities. For governments, it is wasteful in a very direct way to the extent they end up subsidizing costly advanced education of large numbers who will not be able to put it to productive use. For those students encouraged to pursue advanced education for an occupation that turns out to experience excess supply, it is wasteful of many of their potentially most productive years engaged in specialized study that they may be unable to use in the workplace, while older cohorts with such over-supplied capabilities (relative to workforce demand) find themselves exposed to wage suppression and unstable career paths.

Given the importance of avoiding both large shortages and large surpluses of scientists and engineers, there have been many efforts to develop models designed to estimate future paths of employment and wages and to compare these to projections of future supply and demand. This has not however proved to be a successful enterprise, for several reasons.

First, it has proved almost impossible to model future demand for such personnel. According to one informed assessment, future requirements depend on as many as ten forces that are external ("exogenous") to the labor market itself, many of which have proved almost impossible to anticipate. These include inherently unpredictable developments arising from technological change, political decisions, and changing demand for personnel in occupations outside of science and technology. The most predictable of these forces may be those driven by demography, which generally is more slow-moving and predictable than economic or political change, but even demographic change becomes unpredictable if international migration becomes a substantial demographic driver since immigration policies are themselves heavily affected by political and other external decisions. Given these difficulties, most modeling efforts have simply excluded or ignored some of the ten external variables mentioned earlier, with the result that "they can mislead."[8]

Second, the most common form of modeling has been so-called gap analyses, in which separate projection models of supply and demand are developed, and then compared with one another in terms of resulting "gaps." These "gaps" have often been claimed to demonstrate prospective shortages or surpluses of scientists and engineers. Such models fail to allow for the inevitable interactions between supply and demand that result from market mechanisms such as wage adjustments, immigration, transfers of workers across fields, and adjustments in qualifications.

The notoriously misleading "shortfall" models produced in the late 1980s by the NSF's Policy Research and Analysis office (see discussion in chapter 2) provide sobering warnings. These models not only were driven by unwarranted demographic assumptions. They also allowed for no interactions between the supply and demand projections. Finally, they also failed to anticipate at least three powerful exogenous forces: the economic recession of the early 1990s, the end of the Cold War around 1990, and implementation of 1986 legislation that prohibited compulsory retirement (deferred until 1994 for university faculty).[9]

In short, models capable of providing credible forecasts of excesses or insufficiency in the number of scientists and engineers more than a few years into the future have not yet been produced, while models with a variety of weaknesses have been over-interpreted by users apparently unaware of their deficiencies.

How Large Is the U.S. Science and Engineering Workforce?

This is a deceptively simple question that turns out to have several very different answers. Whatever the correct number may be, we can say with assurance that the science and engineering workforce constitutes a very small fraction—most estimates coalesce on about 5 percent—of the overall U.S. workforce of more than 155 million. This percentage, while still quite small, has been growing due to the increasing significance of technical occupations in the economy. In 1983, for example, the comparable percentage for the United States was closer to 3 percent.[10] Similarly small percentages are characteristic of the science and engineering workforces for most countries.

The best statistical overview of the U.S. science and engineering workforce is provided in the National Science Board's magisterial compendium *Science and Engineering Indicators*, which is produced biennially by staff of the National Science Foundation and is based heavily upon NSF data. It offers several disparate estimates that depend upon very different definitions as to who should be counted as a "scientist or engineer." Unfortunately the data available are substantially out of date—on many issues the most recent data are for the year 2008.

The first concept used to define the science and engineering workforce is based on the person's *educational background.* Here, anyone who earned a bachelor's degree or higher in a field of science, engineering, or mathematics is defined as a scientist or engineer, irrespective of whether this person is actually employed in a science or engineering occupation. A second definition focuses on a person's *occupation* rather than on educational background: in this meaning, scientists and engineers are those who are employed in science and engineering occupations, irrespective of whether their educational background is in such fields.

The National Science Foundation provides a useful listing of both degree fields and occupations that it includes within "science and engineering."[11] Some of the large disparities that persist among available estimates of the science and engineering workforce arise from the NSF decision to exclude persons with less than a bachelor's degree. Another source of difference is NSF's inclusion of all social scientists (economics, political science, psychology, sociology, anthropology) as "scientists and engineers" (see discussion in chapter 1).

According to the first "education" definition, in 2008 there were 17.2 million persons in the U.S. workforce who had earned at least one degree, bachelor's and above, in a "science and engineering" field. Of these, 12.6 million had earned their highest degree in these fields, and of these an estimated 4.8 million indicated they were employed in a job closely related to that degree.[12]

According to the "occupation" definition, again for those with bachelor's degrees or higher, there were between 4.8 and 6.4 million scientists and engineers employed in 2008, depending on whether one prefers the data produced by the National Science Foundation or the U.S. Census Bureau.[13]

The Gap between "Educational Background" and "Occupation"

If we compare these two kinds of estimates, it is apparent that a large fraction—on the order of two out of every three—of scientists or engineers as defined by "educational background" are actually employed in occupations that are *not* considered to be in science or engineering. (As noted in chapter 1, Xie and Killewald argue that the fraction is more like one in two if social scientists are excluded.) How might we understand the remarkably large size of this apparent mismatch between those holding degrees and occupations in science and engineering fields? A number of explanations are possible.

First, the disparity may be due to differences in the way educational programs and occupations are classified. Degree fields are defined by a body of knowledge acquired from earning a degree in that field, whereas occupations are defined by the work activities of individuals in that occupation. The National Science Foundation has done its best to define occupations that fall within and outside the rubric of science and engineering. These categories have a long history and are focused on the occupations most closely aligned with degrees in scientific and engineering fields. It would be far more difficult and subjective to determine the extent to which a science or engineering educational background is critical for occupations that are not defined as science and engineering.[14]

A second possible explanation is the (unknown) extent to which large proportions of those with science and engineering degrees are by preference employed outside of science and engineering occupations. The NSF does define a category of "Involuntarily out of field," defined as not working in field of highest degree because a suitable job in field was not available. In general the percentages of respondents placing themselves in this category are very low. A much larger percentage working in occupations outside of science and engineering indicated that their job is "closely or somewhat related to their field," and generally seem to be satisfied with their non-science/

engineering occupation. In other words, it appears that many working outside of these occupations are choosing to do so, based on the opportunities they find available to them. This finding is generally consistent with interpretations that skills acquired in science and engineering degrees are attractive to employers hiring for occupations outside of science and engineering, and indeed that these occupations may offer more attractive careers than do occupations within science and engineering.[15]

A third interpretation is that large numbers of individuals with science and engineering educational background would rather be in a science and engineering occupation but are unable or unwilling to do so because of limited availability of such positions or their unattractive career/employment prospects relative to alternative occupations.

Finally, a fourth explanation is the unknown extent to which the apparently large majority of those with science or engineering degrees but not employed in science or engineering occupations actually are making substantial use of their degrees in their work. On this last issue, the NSF does collect some data that warrant brief attention here, although as we shall see it too does not provide very clear answers. These data come from the 2006 National Survey of Recent College Graduates,[16] mainly from responses to the following survey question:

> To what extent was your work on your principal job related to your highest degree? Was it …
>
> *Mark one answer.*
>
> 1. Closely related
>
> 2. Somewhat related
>
> 3. Not related

The framing of this question severely limits our understanding of how such responses should be interpreted. In particular, consider respondents whose highest degree was in a science or engineering field but are not employed in a science or engineering occupation, and who respond to the survey question that the work on their principal job is "closely related" and/or "somewhat related" to their highest degree. By this they may mean any of the following about their principal job:

> Is closely or somewhat related to technical expertise acquired from their highest degree in science and engineering
>
> Is closely or somewhat related to broader skills they gained as part of their highest degree (e.g., a physics major using quantitative skills in a job as a bond analyst)

Is closely or somewhat related to the need for at least a bachelor's degree, whether or not that degree was in science and engineering.

Yet it is obvious that these three possible meanings among respondents carry with them large differences in interpretation. The National Science Foundation is aware of these ambiguities, has not been satisfied with the analyses they are able to conduct, and is forthright in advising users to be cautious in interpreting these data.[17] Somewhat different survey questions focused on "technical expertise" have been included in other NSF surveys, but unfortunately were not fielded in the 2006 survey, the last for which data have been released.[18]

What Is Meant by "High-tech" Industries and the "High-tech" Workforce?

Much confusion is injected into public discussion of the science and engineering workforce by references to those employed in "high-technology industries" ("high tech" for short). There are differing ways to define this category of industries. The most careful definition is that developed by the U.S. Bureau of Labor Statistics (BLS), which includes those industries with higher-than-average employment of "technology-oriented workers" in the following occupational groups:

Computer and mathematical scientists

Engineers

Drafters, engineering, and mapping technicians

Life scientists

Physical scientists

Life, physical, and social science technicians

Computer and information systems managers

Engineering managers

Natural sciences managers

The BLS then divides industries into three levels of "high tech," depending upon the intensity of their employment of such workers. The fourteen industries in which they constitute at least five times the average of all industries are designated "Level I" high-tech industries and include the following:[19]

Pharmaceutical and medicine manufacturing

Computer and peripheral equipment manufacturing

Communications equipment manufacturing

Semiconductor and other electronic component manufacturing

Navigational, measuring, electromedical, and control instruments manufacturing

Aerospace product and parts manufacturing

Software publishers

Internet publishing and broadcasting

Other telecommunications

Internet service providers and web search portals

Data processing, hosting, and related services

Architectural, engineering, and related services

Computer systems design and related services

Scientific research and development services

Total employment in these Level I high-tech industries in 2002 was about 5.9 million out of total nonfarm employment of 131 million, or 4.5 percent.[20] However, this kind of estimate unfortunately still does not allow any sensible comparison with the estimates of the size of the science/engineering workforce discussed earlier, for the following reason: the workforce estimates for "high-tech" industries encompass *all* employees in such industries—including secretarial, janitorial, administrative, sales, etc.—and not only those in science and engineering occupations.

Are There Generalized "Shortages" or "Surpluses" of Scientists and Engineers?

Given the discussion of the various meanings of "shortages" and "surpluses" previously discussed, what does the evidence show about the existence of general shortages or surpluses, in other words, imbalances between supply and demand, for the broad range of science and engineering occupations?

In a 2011 paper, Philip Martin and Martin Ruhs provide a concise summary statement: "Industries and occupations reporting labor shortages should have rising relative real wages, faster-than-average employment growth, and relatively low and declining unemployment rates."[21]

This is an entirely uncontroversial statement about the way labor markets function in a mixed economy. This perspective has been formally incorporated into the work for the United Kingdom's Migration Advisory Committee, described in greater detail in chapter 8. Economists frequently disagree with one another, but not about this.

It is unfortunate that in the United States there is no continuing, official, and yet independent effort akin to the UK's Migration Advisory Committee that could provide regular objective assessments of claimed labor shortages based on such labor market data. Instead what prevails are recurring rounds of shortage assertions that emanate primarily from employers of scientists and engineers, employer associations, and ad hoc committees established by scientific and business associations (these are discussed in some detail in chapter 4). In the absence of objective sources of ongoing assessment, such claims have been influential and have convinced numerous politicians and journalists that such shortages do indeed exist.

Quantitative Studies

Happily, over the past ten to fifteen years there have been occasional careful studies based on quantitative data, undertaken both by federal labor market analysts and by independent researchers at universities and at independent research institutes such as RAND, the Urban Institute, and the (late) Commission on Professionals in Science and Technology. In addition, there are assessments of career prospects in specific fields that have been undertaken by several professional scientific societies.

Bureau of Labor Statistics (BLS): A significant BLS paper published in 1999 examined labor market data for sixty-eight occupations from 1992 to 1997. This was a period of rapid economic expansion, in which the overall unemployment rate declined sharply from 7.5 percent in 1992 to 4.5 percent in 1998. Notably, this was the same period, described in chapter 2, in which information technology employers and employer associations argued energetically that they faced large and rapidly growing shortages of IT workers that were seriously affecting the industry's ability to expand and maintain its global competitiveness. While the most energetic claims during this period came from IT employers, similar shortage claims were put forward by employers of registered nurses, construction workers, qualified teachers, nannies, and even unskilled workers.[22] In concert with the consensus summarized earlier, the BLS paper noted that if shortages in an occupation were to develop during the Nation's 1991–1998 expansionary period, the occupation's employment growth would be strong; the occupation's wages would increase relative to other occupations, indicating the market response by employers to attract more workers; and the unemployment rate for that occupation would be expected to decline or remain relatively low.[23]

The BLS study applied the following criteria for "shortage" to each of the sixty-eight occupations it examined: employment growth at least 50 percent higher than average employment growth in the period; wage increases at least 30 percent higher than the average rate of wage increase; and an unemployment rate at least 30 percent below the average unemployment rate for all workers.

Notwithstanding the buoyant economic conditions of the period 1992–97, the study identified only seven of the sixty-eight occupations it studied as meeting these criteria for occupations experiencing "shortages." Moreover, these seven did not include information technology, nurses, or the other categories for which employers had been claiming shortages. Those that did meet the quantitative criteria of "shortage" occupations during 1992–97 included only one science or engineering field—mechanical engineers. The other six were: management analysts; advertising and public relations managers; marketing; purchasing agents and buyers; special education teachers; dental hygienists; and airplane pilots and navigators.[24]

The STEM Workforce Data Project: From 2004 to 2007 this project, directed by Richard Ellis[25] and sponsored by the Commission on Professionals in Science and Technology, produced a series of eight data reports on a wide range of subjects, including development of long-term data on STEM employment, degrees, and salaries; the representation in STEM fields of women, minorities, and the foreign-born; and the implications of these data for public policies.[26] The project's final report was completed in 2007, and addresses what it termed the "disconnects" in the America COMPETES Act passed that same year to increase the supply of scientists and engineers in the U.S. workforce:

> [I]t is not clear how plans like these [America COMPETES Act], which are supposed to encourage greater participation in STEM occupations, can overcome the skepticism being displayed on the part of both experienced STEM professionals and young persons about the prospects for these careers ... There is no shortage whatever of interested people. What is in short supply are reasons to believe that technical careers will be worth the considerable investment which they demand in time and training.

To attract more participants, at least two problems require attention. First, at the very time that the nation needs to make STEM careers more attractive, domestic job markets are soft because employers have tapped foreign sources of labor that were not available before.... The second problem is that ... [a]lthough the newly signed America Competes Act claims to respond to the Gathering Storm report and may provide at least some new support for STEM activities, this legislation does not address the labor market conditions that are acting to discourage participation in U.S. science ...

Nevertheless, there are steps that could be taken to improve the outlook for U.S. science and technology. The most obvious move is to address the disconnects in federal STEM policy, which has yet to come to grips with issues like offshoring and the use of guest workers.[27]

RAND Corporation: A series of studies to assess claims of labor shortages in science and engineering have been undertaken since 2000 by RAND's Science and Technology Policy Institute, which for many years was the prime technical advisor to the White House Office of Science and Technology Policy. In a paper presented to a National Academies conference in 2002, six RAND authors noted that "what is meant by 'shortage' has not always been clear," and then considered a set of five different circumstances in which the production of new science and engineering PhDs might be considered too low. Their analyses of available data concluded that

neither earnings patterns nor unemployment patterns indicate an S&E shortage in the data we are able to find ... Altogether the data ... do not portray the kind of vigorous employment and earnings prospect that would be expected to draw increasing numbers of bright and informed young people into S&E fields.[28]

Two of the same RAND authors later helped to organize a 2004 conference focused on data improvements needed for future analyses on this topic, co-sponsored by the White House Office of Science and Technology Policy and the Alfred P. Sloan Foundation. The conference volume pointed to a need for data that:

- account for people trained in STEM fields but working in other occupations;

- focus on STEM labor markets over time and career experience over the life cycle to better understand their responsiveness to changing employment opportunities and earnings.

- provide timely measures of job openings, applications, salaries, graduate student enrollments, and job placements—"flash" indicators like the preliminary estimates of unemployment or inflation that later are adjusted as better but slower-to-emerge data appear— rather than continuing to rely on survey data that are always two to three years out of date.

• address the demand side of these labor markets, including job slots, job offers, facilities and equipment that complement or substitute for labor, demand for products and services produced, etc.

• provide insight into how employers adjust to occupational labor shortages, including evidence on wages, recruitment efforts, working conditions such as overtime, modification of educational requirements for new hires, increased use of immigrants, or use of contracting out or offshore outsourcing.

• facilitate more accurate forecasts of conditions in these labor markets.[29]

In 2008, two other researchers at RAND's National Defense Research Institute produced a report at the request of the Under Secretary of Defense for Personnel and Readiness, addressing the adequacy of the U.S. science and engineering workforce in general for maintaining its position of leadership in science and technology, as well as the special requirements of the Department of Defense. It concluded: "In sum, unemployment and wage growth patterns are thus not unusual and do not point to the presence of a chronic or cyclical shortage in S&E."[30]

As to the puzzle surrounding greater growth in S&E employment as compared with degree production in the United States, they noted:

The much higher rate of increase of occupational employment in S&E compared with degree production suggests two explanations. First, obtaining a degree in S&E is only one of several paths to joining the S&E workforce. This is consistent with the influx of S&E workers either from abroad or from non-S&E occupations ...

Second, taken together, the high S&E employment growth, the slower S&E degree growth, and the high fraction of degrees awarded to foreign-born graduate students suggests that many U.S.-born graduate students are choosing to study non-S&E fields, perhaps because the job opportunities, challenges, and earnings are perceived to be greater there than in S&E occupations or because U.S. citizens experience more competition from foreigners for spots at science and engineering colleges and universities.[31]

As to the general question of whether the United States is falling behind other nations in science and technology, the report concluded that the United States is still a premier performer in science and engineering and in many measures of such prowess improved faster than did Japan and Europe. Developing nations such as China, India, and South Korea, though starting from a small base, showed more rapid growth in science and engineering, and if this were to continue the United States should expect its share of world science and engineering output to diminish.[32]

University-based Research

Over the past ten to fifteen years a substantial body of research has been conducted by independent researchers, mostly university-based labor economists and other social scientists, on supply and demand dynamics in the U.S. science and engineering workforce. This work has begun to plow what previously had been a rather fallow field of basic research and analysis about these critically important but relatively small parts of the U.S. workforce.

As might be expected, the results of this academic research has been more sophisticated and nuanced than the literature produced by employer associations and others with strong interests in the outcome. It has also been characterized by a much wider range of findings and conclusions; while a common theme of "shortages" has tended to emerge from the employer-supported analyses, the findings of academic research have been far more diverse. Of course in this brief discussion we cannot do justice to this now very substantial body of serious research, but let us touch briefly upon some of the highlights.

Among the leaders of this burgeoning academic research area has been Richard Freeman, professor of economics at Harvard and for many years director of the Labor Studies program of the National Bureau of Economic Research (NBER).[33] With financial support from the Sloan Foundation, the National Science Foundation (NSF), the National Institute of Standards and Technology (NIST), the National Institutes of Health (NIH), and other funders of science and engineering, Freeman and his associates have produced a large volume of analytic work and quantitative analyses related to the U.S. science and engineering workforce.

In a wide-ranging chapter published in a 2010 NBER book,[34] Freeman catalogs the massive spread of university education in the latter part of the twentieth century and notes in particular that the global numbers studying science and engineering had increased almost 400 percent between 1970 and 2006. The already-large numbers in U.S. institutions did increase during this period, but most of the global increase was due to rapid expansion of higher education in Asia and Europe.[35] Whereas the United States in 1992 awarded some 23 percent of first degrees in all subject areas awarded in the key regions of North America, Europe, and Asia combined, the comparable U.S. percentage for 2004 was reduced to 15 percent because of the rapid expansion under way over that period in Europe and Asia. For first degrees in natural science and engineering fields, the comparable change for the United States was from 13 percent in 1992 to 10 percent in 2004.[36]

At the PhD level for science and engineering fields, the U.S. share fell from about 22 percent in 2001 to 18 percent in 2004. Once again the U.S. numbers increased over this period, but the increases in China were far more rapid, as were those in Europe. Freeman estimated that, given the very rapid rate of expansion in Chinese higher education and the country's enormous base population (well over four times larger than the U.S. population), it would likely produce more PhDs in science and engineering than the United States by around 2010, although the quality of such degrees would likely be questionable. In addition, a substantial fraction of such U.S. degrees would be earned by international students, especially from China.[37]

The effects of these massive increases in higher education in China are mixed; they had expanded educational opportunities, but also had led to major problems of unemployment and underemployment for graduates in the Chinese labor market, even before the onset of the global recession in 2008. Freeman also cites 2005 survey data developed by McKinsey Global Institute showing that multinational recruiters considered only 10 percent of engineering graduates from China and 25 percent from India would meet their requirements in terms of skills, language, and potential mobility, although these data did not explore whether their expected remuneration would be sufficiently lower to compensate for their lower qualifications.[38]

In an earlier paper, Freeman concludes that the job market in the United States had been deteriorating for younger workers in science and engineering, thereby discouraging U.S. students from pursuing careers in these fields. Yet the rewards of this deteriorating job market were still sufficient to attract large numbers of migrants from developing countries in which career and earning prospects are even lower.[39]

Freeman points to countervailing impacts of these trends upon the U.S. economy and its science and engineering workforce. On the one hand, the expanding number of highly educated scientists and engineers around the world should accelerate scientific and technological advances and the positive economic effects they can produce, and also reduce the cost to U.S. consumers of the goods they export to the United States by raising their economic productivity. On the other hand, it should also enhance those countries' abilities to compete effectively with U.S. producers in high-tech sectors in which the United States has long held a comparative advantage, thereby reducing career opportunities and earnings in these important parts of the U.S. workforce and economy.[40]

Another Harvard labor economist, George Borjas, has examined the impact of high-skill immigration on U.S. labor markets, especially at the doctoral level. Using data from the NSF's Survey of Earned Doctorates and Survey of Doctoral Recipients, his findings broadly are that these high-skill labor markets operate like others. The "supply shock" of rapid increases in the number of foreign-born doctorates has "a significant and adverse effect on the earnings of doctorates who graduated at roughly the same time," regardless of whether they are native- born or foreign-born. In quantitative terms, he estimates that a 10 percent immigration-induced increase in the supply of doctorates lowers the wage of competing workers by about 3–4 percent. About half of this adverse wage effect can be attributed to the increased prevalence of low-pay postdoctoral appointments in fields that have softer labor market conditions because of large-scale immigration.[41]

Paula Stephan, another leading researcher on science and engineering also associated with the NBER group, published an impressive book in 2012 entitled *How Economics Shapes Science*, that includes the following conclusions among others:

- "Predictions of shortages of scientists and engineers occur with some frequency, despite evidence to the contrary … [and] have often strayed considerably from the underlying reality."[42]

- Labor market conditions do affect career decisions by potential new entrants: there is "considerable evidence that the number of individuals choosing to follow a course of study in science and engineering is responsive to market signals" such as career prospects, rates of unemployment and remuneration, and that such signals can indeed be affected by the volume and pace of skilled immigration.[43] As one example she summarizes research on the quasi-experiment provided by U.S. labor market trends in mathematics during the 1990s, after the collapse of the USSR enabled large numbers of Soviet mathematicians to emigrate to the United States. The data show that academic remuneration in the field declined by 8 percent (adjusted for inflation) and unemployment rates increased. Increasing fractions of new PhDs found themselves consigned to temporary positions: the number of tenure-track faculty in traditional mathematics departments declined by 27 percent, while that for non-tenure-track faculty increased by 37 percent. In response to these "dismal job prospects," the number of U.S. students applying for entry to graduate mathematics programs in the mid-1990s decreased by 30 percent.[44]

- "Salaries for PhDs in science and engineering have been relatively low for a substantial period of time" and the amount of time required for doctoral (and often postdoctoral) training supported by low stipends exacts another large financial cost on lifetime earnings relative to alternative career paths available to talented students. She calculates alternative lifetime earnings (at present value) for a hypothetical student who chooses to pursue either a PhD in science (biological, mathematical, or statistics) or an MBA at the same university, and finds that the science path would produce only about 67 percent of the earnings of the MBA path: about $2.2 million for the former versus $3.2 million for the latter. Her hypothetical calculations are conservative in that they assume that the student choosing the PhD path would receive full financial support while the costs of the MBA degree would be borne entirely by the student, and they also assume that the hypothetical MBA would not receive any of the bonuses or stock options that are common in that career path.

•   For engineering, in which most career entry is at the bachelor's level, she finds similar responsiveness to market signals. The fraction of engineering degrees among all bachelor's degrees in a given year "closely tracks the career prospects of engineers four years earlier—when the students were freshmen—as measured either by the present value of earnings in engineering relative to other occupations" or by the relative wages in engineering at the time the students entered college.[45]

In a series of papers since 2007, Giovanni Peri of the University of California at Davis has published a strongly positive set of judgments about the economic effects of immigration by scientists and engineers (he also judges low-skill immigration to be quite positive, but this is not relevant for our discussion). The thrust of his argument is that because scientists and engineers who migrate to the United States (on both permanent and temporary visas) have made valuable contributions to the engines of economic growth, admission of more such migrants will further increase U.S. economic growth and well-being. Unlike some of the other researchers discussed in this section, Peri does not seriously address conditions in the U.S. labor market for scientists and engineers, nor possible impacts these may have on the attractiveness of these careers for U.S. citizens. Instead he argues that the positive effects of highly skilled migrants to U.S. economic growth improve productivity and wages for everyone.

In a paper commissioned by the Hamilton Project (a Brookings Institution effort co-founded by Robert Rubin), Peri argues for a multiphased reform of the immigration system that would begin by expanding the numbers of temporary visas for scientists and engineers. All such visas would be allocated by auction to the highest bidding employers. A similar idea was proposed by Pia Orrenius and Madeline Zavodny in 2010; many economists are attracted to auction systems as efficient indicators of market forces.

In phase 2 of Peri's proposed reform, current permanent employment-based visas would be transformed into "provisional" visas, also sold to the highest-bidding employers at auction and good for five years, after which all such visa-holders who met minimal standards of employment and legal compliance would be able to apply for permanent residence visas. In the final phase 3, some of the permanent immigration visas now provided to extended-family preferences and "diversity" visas would be shifted to these employment-based categories, and all foreign students graduating from U.S. universities with a job offer would be automatically eligible for such visas.[46]

In their 2011 study, Anthony Carnevale and his associates at Georgetown University forecast that future demand will be rising strongly for what they call "STEM-capable" workers both in traditional STEM occupations and in other non-STEM occupations for which similar competencies will be needed. They seek to resolve the "shortage versus surplus puzzle" by reference to "diversion" of STEM-capable workers into non-STEM occupations.[47] They agree with most other researchers that the U.S. education system is in fact producing more than enough graduates to fill available positions in STEM occupations, but believe that many STEM-capable workers pursue more attractive career paths in non-STEM occupations—especially managerial, professional, and healthcare occupations—for which the quantitative skills and cognitive abilities of STEM graduates are being more highly valued and hence potential earnings are higher and more rapidly increasing.[48] As Carnevale stated to the *Wall Street Journal*, "If you're a high math student in America, from a purely economic point of view, it's crazy to go into STEM [occupations]."[49]

They depart from many others who concentrate on the STEM workforce at the graduate and postgraduate levels, concentrating instead on those at the sub-baccalaureate level who have earned postsecondary certificates, some college credit, or Associate's degrees. They argue that the U.S. system does well in producing or attracting the highest performing STEM workers at the baccalaureate level and above, but performs less well in producing STEM sub-baccalaureate personnel who are needed for essential occupations such as engineering technicians, systems administrators, and related functions. For this reason they emphasize the need to address weaknesses at the level of U.S. community colleges rather than at bachelor's and higher levels. In particular, they note that current funding structures that support community college education tend to favor less costly "general" rather than "technical" programs, especially as financial resources per student have been cut.[50]

As noted in chapter 1, a 2007 article by Lowell and Salzman of Georgetown and Rutgers Universities specifically assessed the arguments and data undergirding the National Academies' *Rising Above the Gathering Storm* report released in 2005. In so doing they addressed many of the issues concerning labor shortages of scientists and engineers. Like other independent analysts, they were unable to find empirical evidence of such shortages in any of the data they examined. Here is a concise summary of their findings:

> [A]ssessing the claims of labor market shortages is crucial. Purported labor market shortages for scientists and engineers are anecdotal and also not supported by the available evidence. Little analysis has been conducted of hiring difficulties by firms and the supply of workers. A particular employer or industry's experiences in hiring could be the result of any number of factors. The assumption that difficulties in hiring is due just to supply can have counterproductive consequences: an increase in supply that leads to high unemployment, lowered wages and decline in working conditions will have the long-term effect of weakening future supply by discouraging current students.[51]

A more recent book by Yu Xie and Alexandra Killewald[52] at the University of Michigan, published in 2012 and mentioned briefly in chapter 1, analyzes an even wider range of relevant datasets. The book is focused on the sciences subset of the science and engineering workforce. Overall their conclusions were quite consistent with those of earlier analysts. Analyses of relative earnings provided no evidence of "shortages" of scientists in the U.S. workforce:

> [T]he results show that scientists have not fared well in earnings in the past five decades. Compared with other elite professions we studied, the relative earnings of basic scientists have either lost ground or stagnated. More specifically, the earnings of basic scientists have declined relative to those of engineers, computer scientists, nurses, medical doctors, and lawyers.[53]

Actually, Xie and Killewald's comparisons of earning trends for basic scientists as compared with lawyers may understate the extent to which the scientists have lost ground. They are comparing earnings of scientists with PhDs that take six or more years with those of lawyers whose degree takes only three years—if the comparison had been to scientists with master's degrees, which take about two years, the earnings advantage of lawyers might have been even larger.

Of course people choose careers for reasons other than economic rewards. Outstanding scientists have the opportunity to make truly revolutionary discoveries that can enhance human well-being or understanding and lead to distinguished research positions and even worldwide acclaim. Xie and Killewald argue that the enormous expansion of U.S. science has made this a high-risk proposition, however, since only a tiny handful can achieve the status of great scientist, Nobel Laureate, etc. The rewards of stellar research success may be high, but the prospects are low—almost like a lottery—which they argue may lead talented individuals to pursue careers that have more predictable paths and more attractive economic rewards.[54]

They also agree with others' findings that U.S. institutions are awarding far more degrees in science and engineering than are employed in science and engineering occupations. However, they argue that the ratio has been overestimated in the National Science Board's *Science and Engineering Indicators* data, due to the decision to include the social sciences in the science and engineering category. If degrees in the social sciences are excluded, Xie and Killewald estimate that about half of science and engineering degree-holders are employed in science and engineering occupations, versus the one-third estimated when social science degrees are included.[55]

Their answer to the question posed by their book title—*Is American Science in Decline?*—is that it is not. To the contrary they describe U.S. science as continuing to be globally dominant in most fields, and continuing to strengthen. At the same time, the large gap between the United States and other countries is narrowing as others are improving more rapidly, while the United States experiences what they call "saturation."[56] If these narrowing trends continue over the longer term, U.S. science may lose the globally dominant position it currently holds even if it continues to increase in size and quality. They emphasize that "Loss of dominance, however, is not the same thing as decline."

Are There Particular Fields, Industries, Categories of Employers, or Geographical Regions for Which Supply and Demand Are Not in Balance?

While most objective analysts have not been able to find empirical evidence of U.S. demand in excess of supply for scientists and engineers in general, it is nonetheless possible that this can happen in certain subfields of science and engineering, in certain time periods, among certain categories of employers, or in certain geographical regions. Consider the following.

New subfields: New subfields or technologies that experience rapid growth can quickly outstrip the necessarily limited supply of experts. Evidence of strong demand for such experts can evoke a strong supply response as students and professionals shift their focus, but such supply responses often are slower if demand rises very rapidly, generating temporary "shortages." The booming demand during the 1990s for experts on the Internet and biotechnology provides obvious examples, although in both cases boom turned to bust in less than a decade.

Sudden expansion: Fields that experience rapid growth for reasons unrelated to the science or technology itself can generate equally rapid acceleration in

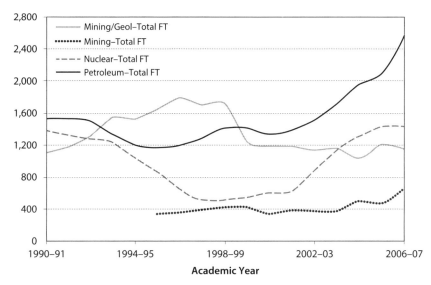

Figure 5.1.

Undergraduate enrollments, total full-time (FT) and first-year, in energy-related engineering specialities, fall of academic years 1990–1991 to 2006–2007.

Source: "A Comprehensive Statistical Study of Education Trends in Three Disciplines Closely Linked to Energy-Producing Industries—Mining, Nuclear and Petroleum Engineering," Engineering Trends, http://engtrends.com/IEE/1107C.php.

demand for professional workers. A clear example here is petroleum engineering. As noted in chapter 3 (see figure 3.1), the real price of petroleum rose rapidly during the 1970s due to supply disruptions, then declined sharply until the second half of the 1980s and remained relatively stagnant for another decade-and-a-half. During this extended period, there was little incentive to invest in exploration and development of relatively high-cost sources such as those in North America, and hence demand for petroleum engineers in the United States and Canada stagnated. Enrollments in petroleum engineering degree programs also declined and stagnated (see figure 5.1), and as the enrollments declined so too did the number of engineering schools offering programs in petroleum engineering—from twenty-four to twenty between academic year 1985–86 and academic year 2005–06. Conversely, the number of bachelor's degrees awarded continued to rise for several years because of the time required for an entering freshman to complete the degree.[57]

Then, around the turn of the twenty-first century, world oil prices again began to rise very sharply, suddenly making energy prospecting and development in North America economically attractive. Meanwhile new techniques for extraction of shale oil and gas had been developed. Together these developments generated rapid growth in both demand and remuneration for petroleum (and related natural gas) engineers. As depicted in figure 5.1, subsequent enrollment in petroleum engineering programs rose rapidly in response to rising demand, increasing from around 1,400 full-time undergraduate enrollees in 2000 to 2,600 in 2006–07. First-year enrollments rose even faster in proportional terms, from about four hundred to about eight hundred over the same short span of years. Similar, if smaller, increases occurred in nuclear engineering.

Government-financed Initiatives

Demand trends for some science and engineering fields are heavily affected by government expenditures. An obvious example here is the large numbers of scientists and engineers employed in the successful but time-limited Apollo mission during the 1960s, as discussed in chapter 2. Other examples include the defense buildup of the 1980s, and the creation of the Department of Homeland Security and expansion of other domestic and international security expenditures that followed the 2001 attacks on New York and Washington. (The political decision to double the NIH budget from 1998 to 2003 also increased demand for biomedical researchers. However, because the increased funds were allocated primarily to university-based research, the increase in effective demand was expressed primarily for additional graduate and postdoctoral research assistants and non-tenure track faculty, rather than for permanent faculty and research employees.)

Geographical Concentrations

The classic case here is Silicon Valley, which has experienced a rapid-fire series of booms and busts in demand, driven by the rise and fall of firms in the computer, software, and information technology industries that are heavily concentrated in this small geographical region. There are numerous and credible anecdotes that during boom times in the Valley (e.g., from around 1997 to 2001), such firms were offering substantial hiring bonuses to attract personnel from competitor firms. While there may have been large numbers of comparable professionals who could be recruited from elsewhere in the country, the high housing prices in the Valley limited such internal migration from regions with far lower housing costs. Then, when Silicon Valley boom turned to Silicon Valley bust in 2001, thousands of professionals experienced career crises as their employers failed or laid off substantial fractions of their Valley workforces.

Exceptionally high housing prices in Silicon Valley continue to hinder employers' efforts to recruit experienced personnel from lower cost areas of the United States, unless employers are willing to offer remuneration high enough to equalize such price differentials. Most Silicon Valley employers appear to be unwilling to do so, preferring instead to recruit domestically from pools of recent graduates and internationally from countries with far lower income levels.

The Significance of Trends and Rates in PhD Production

It is common in some of the reports described in this book to assess the contribution of a country's higher education system to its economy in part by the number of PhDs awarded in science and engineering. The *Rising Above the Gathering Storm* report, for example, pointed forcefully to the growing competition in PhD production from countries such as China and South Korea:

> One measure of other nations' successful adaptation of the US model is doctoral production, which increased rapidly around the world but most notably in China and South Korea … In South Korea, doctorate production rose from 128 in 1975 to 2,865 in 2001. In China, doctorate production was essentially zero until 1985, but 15 years later, 7,304 doctorates were conferred. In 1975, the United States conferred 59% of the world's doctoral degrees in science and engineering; by 2001, our share had fallen to 41%. China's 2001 portion was 12%.[58]

There are two critical but often unstated assumptions that underlie this kind of assessment. First, there is the assumption that the PhD is the most critical degree in science and engineering fields, and that increasing the numbers of such PhDs will increase leading-edge basic and applied research which in turn will produce more rapid growth in the country's economy. Second, there is the assumption that the designation "doctorate" or "PhD" represents a homogeneous category of postgraduate education for which production numbers can be usefully compared cross-nationally.

As to the first, the realities are that PhDs represent only small fractions of any nation's workforce in the economically dynamic fields of engineering and computing. In the United States, most engineers become qualified professionals with no more than a bachelor's degree, and the master's degree is considered the gold standard for engineering graduate education. In 2009, for example, the number of master's degrees earned in engineering fields equaled 52 percent of the number of bachelor's degrees earned, whereas the number of PhDs earned was only about 11 percent. Calculated another way, the PhD represents a considerably smaller proportion of graduate degrees in engineering (about 18 percent) than in the natural sciences (about 36 percent).[59]

The number of PhDs earned in both engineering and science is small compared with the number of master's in part because the earnings premium for the PhD is relatively small. Some industries (notably biotechnology) do employ higher proportions of PhDs in their R&D workforce than do engineering-focused industries. Yet even in biotechnology the majority of research and development positions are filled by those with either bachelor's or master's degrees.[60]

The U.S. Department of Defense, one of the largest employers of civilian scientists and engineers, reports that (as of 2011) approximately half of its STEM workforce had earned no more than a bachelor's degree. Nearly one-quarter held master's degrees, and only about 5 percent had a doctoral degree. Indeed, in the important field of the computer and mathematical sciences—one of the largest of the STEM fields so central to both the U.S. military's leading-edge technical capabilities and to concerns about the United States falling behind its competitors—nearly half of that department's civilian STEM workforce actually has *less* than a bachelor's degree.[61]

As to the second assumption, it is well known that PhD or doctoral degrees are by no means homogenous cross-nationally. The greatest increase in PhD production in recent years has been that produced by the burgeoning universities of China, which by one count produced some 50,000 PhDs in all disciplines in 2009. Yet, observers both within and outside China acknowledge that quality control on these degrees is weak—Chinese PhDs take only three years to complete, the PhD supervisors are themselves often not well qualified, and poor students are not weeded out. Most PhD graduates from Chinese universities are not competitive for positions outside of China that require doctoral degrees, though the rapid expansion of Chinese higher education makes it relatively easy for them to find academic positions within China.[62]

There is also wide variation in the quality of the PhD even within a given country. In the United States there are few national or state-level standards for PhD degrees in most fields, unlike degrees in medicine and law. Indeed, the National Research Council in the United States expends considerable effort in assessing the relative quality of U.S. PhD programs across all disciplines.[63]

In short the wide quality range of PhDs is well-known by academics around the world, though perhaps less so by those politicians who urge their countries' universities to expand the numbers of PhDs they produce as a critical driver of future economic prosperity.[64] The popular expedient of comparing the numbers and trends of PhDs produced across countries may well disguise more than it reveals.

How Does the Number of Recent PhDs Compare with PhD Career Opportunities in the United States?

The trends and international comparisons of PhD production focus solely on the supply side of the PhD degree, whereas the reality is that there also must be effective demand if those educated to this advanced level are actually to conduct the research for which the PhD is designed to prepare them. Specifically, employment positions must be available for such PhDs; funding to support their research activities must be available; and in most fields they must have access to the laboratory facilities and equipment that are required for them to conduct research. In addition, before such research can contribute to a country's economic growth and prosperity, there also must be effective demand for any products or patents that might eventually emerge from such research.

Unless a country is producing large fractions of its PhDs for export (as the Philippines does for nursing and physical therapy degrees at all educational levels, for example), the next question is to compare the annual increment to science/engineering PhDs in a country with the number of open positions in its labor market for doctoral holders in these fields. There is a substantial research literature on this topic, which generally reports:

- Wide variation across science and engineering fields;

- Significant variations in the proportion of new PhDs finding employment in universities versus non-academic settings, and in career-path versus temporary and often poorly paid positions such as postdocs.

- Substantial shifts in the supply/demand situation from year to year, depending on the business cycle, availability of governmental research funding, and the financial condition of institutions of higher education.

Labor economists normally would assess the relationship between supply and demand for such a category of workers on the basis of remuneration levels and trends. As noted earlier, by this standard the number of PhDs in most (not all) U.S. fields of science and mathematics seems to be in ample supply —some would say "excessive" supply—relative to demand.

Compared to other occupations requiring many years of education beyond the bachelor's, remuneration levels for PhDs in most (but not all) fields of science are relatively low, and especially so in relation to the investments of time and money required to earn the degree. Data for 2006–08 from the American Community Survey show that MDs have earnings that are fully 102 percent higher than PhDs in the biological, mathematical, and physical sciences. Lawyers, whose professional degree takes only three years, have earnings 52 percent higher than science PhDs, and even PhD social scientists show a 13 percent earnings premium. Engineers and computer scientists with PhDs earn more than do those in the natural sciences, with premiums of about 22–23 percent.[65]

Moreover, in some PhD fields in the natural sciences—and especially in the largest category, the basic biomedical sciences—the length of time-to-PhD is quite long. The most recent data show an average time-to-degree of more than 6.5 years, at which point the average biomedical PhD recipient is over thirty years of age.[66] Even then most find that they must undertake a further "postdoctoral" period of two to five years in order to be competitive in the difficult biomedical labor market. The "opportunity costs"—that is, the foregone earnings during these extended periods of student and quasi-student statuses—are very substantial, as described in the earlier summary of work by Paula Stephan and in other work.[67] Finally, while the PhD has long been seen as a research degree designed for those wishing to pursue research in academe, the majority of recent PhDs seem instead to be finding employment outside of academe.[68]

All of these trends have been amply reported in the specialist journals and magazines focused on higher education, science, and engineering. In the summary words of a late 2009 story in the *Chronicle of Higher Education*:[69]

> It is the best of times and worst of times to start a science career in the United States.
>
> Researchers today have access to powerful new tools and techniques—such as rapid gene sequencers and giant telescopes—that have accelerated the pace of discovery beyond the imagination of previous generations.
>
> But for many of today's graduate students, the future could not look much bleaker.
>
> They see long periods of training, a shortage of academic jobs, and intense competition for research grants looming ahead of them. "They get a sense that this is a really frustrating career path," says Thomas R. Insel, director of the National Institute of Mental Health ... some of the brightest students in the country are demoralized and bypassing scientific careers ...
>
> Despite the realities on the ground, leaders at the top of government, academe, and industry insist that the nation needs more scientists.... "Rising Above the Gathering Storm" ... acknowledged that "the recommendations for additional support for thousands of undergraduates and graduates could be setting those students up for jobs that might not exist." But it dismissed such fears with the vague proclamation that the number of doctorates "has not kept pace with the increasing importance of science and technology to the nation's prosperity."

A 2011 special report published in *Nature* magazine on the PhD in international comparative terms concluded:

> In some countries, including the United States and Japan, people who have trained at great length and expense to be researchers confront a dwindling number of academic jobs, and an industrial sector unable to take up the slack. Supply has outstripped demand and, although few PhD holders end up unemployed, it is not clear that spending years securing this high-level qualification is worth it for a job as, for example a high-school teacher.... To Paula Stephan, an economist ... who studies PhD trends, it is "scandalous" that U.S. politicians continue to speak of a PhD shortage.[70]

An editorial accompanying *Nature's* 2011 report concluded that in the United States and in Japan,

> Exceptionally bright science PhD holders from elite academic institutions are slogging through five or ten years of poorly paid postdoctoral studies, slowly becoming disillusioned by the ruthless and often fruitless fight for a permanent academic position. That is because increased government research funding from the US National Institutes of Health and Japan's science and education ministry has driven expansion of doctoral and postdoctoral education—without giving enough thought to how the labour market will accommodate those who emerge. The system is driven by the supply of research funding, not the demand of the job market.[71]

Such perspectives, which appear frequently in media focused on scientific research and education, are surprisingly rarely mentioned in political debates and mainstream media treatments about the STEM workforce.

Concerns about PhD Career Paths Recognized by NIH Leadership

To their credit, the senior leadership of the National Institutes of Health (NIH) has recognized these issues and has been seeking strategies for addressing them. In 2011, the director of NIH Dr. Francis Collins, appointed a high-level Biomedical Research Workforce Working Group of his Advisory Committee to the Director. The Working Group was co-chaired by Dr. Shirley Tilghman, president of Princeton University, and Dr. Sally Rockey, NIH deputy director for Extramural Research. Its charge was to:

> Develop a model for a sustainable and diverse U.S. biomedical research workforce that can inform decisions about training of the

optimal number of people for the appropriate types of positions that will advance science and promote health. Developing the model will include an analysis of the current composition and size of the workforce to understand the consequences of current funding policies on the research framework.[72]

The full report of this working group well repays review by those interested in the current and prospective quality of the PhD biomedical research workforce in the United States. In brief summary, the working group concluded that:

The combination of the large upsurge in U.S.-trained PhDs, increased influx of foreign-trained PhDs, and aging of the academic biomedical research workforce makes launching a traditional, independent, academic research career increasingly difficult.

The long training time and relatively low early-career salaries when compared to other scientific disciplines and professional careers may make the biomedical research career less attractive to the best and brightest of our young people.

The current training programs do little to prepare people for anything besides an academic research career, despite clear evidence that a declining percentage of graduates find such positions in the future.[73]

During the oral presentation of the Working Group report to the Advisory Committee to the Director, both Dr. Tilghman and Dr. Collins summarized the current system by which the biomedical research workforce is educated and deployed as "dysfunctional." Both stated that the NIH has a responsibility to examine possible adjustments that can improve its quality and sustainability going forward, while taking care to avoid doing unintended harm to the high productivity of the research system.[74] It remains to be seen whether the analysis and advice from this working group will lead to any of its recommendations being adopted by the NIH and the broader biomedical research community.

How Many H-1B and Other Temporary Visa-Holders Work in the U.S. Science and Engineering Workforce?

As discussed in another context in chapter 4, substantial numbers of foreign-born scientists and engineers enter the U.S. labor market each year under an array of "temporary" visas. These visas are "temporary" in the sense that they do not provide for permanent residence, yet many such visas allow multiple years of legal residence and employment. The H-1B visa has become one of the most controversial aspects of current U.S. immigration policy. This is hardly the place for a discussion of these controversies, but it is appropriate to point to the gaps in our knowledge about such "temporary" visas.

We do know that the annual issuance of new H-1B visas is quite large, substantially exceeding 100,000 in recent years. We also know that H-1B visas are effectively valid for six to seven years or more, and hence that the "stock" of H-1B visa-holders resident in the United States is considerably larger than the annual flow. Surprisingly the U.S. government currently does not make available any estimates on the size of this accumulated "stock" of H-1B visa-holders currently working in the United States, nor does it enable others who might develop such estimates to obtain the raw data they would need. Estimates by independent researchers, based upon the limited data they are able to obtain from the government, are the best we have. Yet these estimates are substantially out of date due to unavailability of more recent raw data.

The most credible and widely cited estimates suggest that the H-1B "stock" population doubled from around 1996 to 2003, and by 2005 had accumulated to over 500,000.[75] The year 2005 is the last year for which the necessary data to make such estimates have been released by the U.S. government. Most but not all of these are in science and engineering fields. If the figure of "over 500,000" for 2005 is at all close to reality, such "temporary" visas may represent a substantial fraction of total number in all science and engineering occupations (estimated by the NSF to be about 5 million in 2006) and an even higher fraction of those employed in the computer-related and IT fields in which H-1B visas have been concentrated.

There are many other temporary visa categories in addition to the H-1B. These are designated in legislation with a letter from the alphabet, and hence sometimes referred to as an "alphabet soup" of temporary visas. Because the number of such visa categories greatly exceeds the twenty-six letters in the English alphabet—there are now well over seventy such visa classes—many are broken into subcategories using numbers and sometimes additional letters as well. This is the origin of the now-famous "H-1B" visa described earlier. Examples include:

- F-1 (student): some holders of F-1 visas are employed twenty hours per week and full-time during vacations

- J-1 (exchange visitor): many are used for postdocs employed full-time as research assistants

- L-1 (intracompany transferee), and L-2 (dependents of L-1): spouses are permitted to work in the United States

- O-1 ("person of extraordinary ability"): numbers are small

- TN (for Canadian/Mexican nationals): provide "professional services" for employers

- E-3 (similar to H-1B, but for Australian nationals): two-year visa, extendable indefinitely

As is the case for the H-1B visa, data for these other temporary visas are also quite weak. There has been a vigorous though often chaotic and ill-informed debate about the pros and cons of this set of visa policies. In the absence of credible data on the total size of the "stocks" of scientists and engineers holding all such temporary visas, this debate is heavily influenced by advocacy and anecdote rather than empirical evidence.

How Many Postdocs Are There in the United States?

While this also might sound like a rather simple question—one might imagine that the relatively small number of U.S. research universities that host large numbers of postdocs could easily provide annual counts of their postdoctoral positions—in reality the available data preclude a clear answer. Postdoc data are fragmentary, and definitional problems abound—one thoughtful early report on the topic concluded that different universities and departments use at least one dozen distinct occupational titles that all might fall under the postdoc rubric.[76] After many years of discussion, the National Institutes of Health (NIH) and the National Science Foundation (NSF) in collaboration with the National Postdoctoral Association agreed in 2007 upon the following definition of a postdoc:

> An individual who has received a doctoral degree (or equivalent) and is engaged in a temporary and defined period of mentored advanced training to enhance the professional skills and research independence needed to pursue his or her chosen career path.[77]

The authors of the 2008 *Science and Engineering Indicators* usefully attempted to piece together an estimate of the total number of postdocs from existing sources, including the Survey of Doctoral Recipients (SDR) and the Survey of Graduate Students and Postdoctorates in Science and Engineering (GSS). They acknowledged that the results of their efforts were incomplete and fragmentary. By design the SDR pool is limited only to postdocs who received their doctorates from U.S. universities, in other words it excludes all postdocs whose PhDs were awarded by non-U.S. institutions, a category of postdocs that appears to have become very large. Meanwhile, the

GSS estimate—based upon counts of postdocs provided by graduate science, engineering, and related academic departments—does include postdocs whose doctorates were earned abroad, but excludes postdocs in non-academic positions or in units that do not offer graduate degrees, thereby missing evidently large numbers of postdocs working at nonacademic research centers operated by government agencies, corporations, and nonprofit organizations. Both the SDR and the GSS surveys exclude foreign-educated postdocs in non-academic employment, and suffer from the uncertainties within universities as to which occupational titles should be included under "postdocs."[78]

Acknowledging these substantial limitations, *Science and Engineering Indicators* estimates that on the order of 89,000 postdocs were employed in the United States as of late 2005, representing a large increase from prior years. The estimate of 89,000 is composed of the following categories:

- 22,900 who are U.S. citizens or hold permanent resident visas in academic postdoc positions (SDR estimate).

- 26,600 on temporary visas who are in academic postdoc positions (GSS estimate).

- 13,000 U.S.-educated persons who are occupied in postdoc positions not covered by GSS (SDR estimate).

- 26,500 postdocs who are on temporary visas and in positions not covered by GSS (estimate derived by assuming that the proportion of temporary-visa postdocs in other sectors and other parts of academia is the same as in the portion covered by GSS).

While this gross estimate provides useful insights as to the scale of the postdoc population, the *Science and Engineering Indicators* report states explicitly that "other, comparably plausible assumptions lead to a substantially different total." Moreover, it should quickly be reiterated that this gross estimate, while the most current available, is based upon data from late 2005 and hence is itself considerably out of date.[79]

How Do Remuneration Trends in Science and Engineering Compare with Those in Other Fields?

The remuneration picture for U.S. scientists and engineers is a mixed one, and cannot be summarized in a sentence or two.

First, those who earn degrees in science and engineering fields generally have higher remuneration than those with comparable-level degrees in arts, humanities, or the social sciences (except in economics).[80] However, data from the U.S. Census and the American Community Survey show that natural scientists (combining biological, mathematical, and physical sciences) generally have lower earnings than do engineers, computer scientists, and social scientists—these differences are not small, on the order of 20–30 percent lower across different levels of degree.

Yet as noted earlier, the largest earnings differences appear to be at the doctoral level, most notably when scientists and engineers (all fields) are compared with both physicians and lawyers. Physicians' earnings in 2006–08 exceeded those of natural sciences PhDs by about 100 percent, and lawyers' earnings exceeded the scientists' earnings by about 50 percent. Moreover, these disparities have risen substantially since 1960, and especially since 1990.[81]

Second, within science and engineering there is much variation in remuneration by fields. Here a great deal depends on the extent to which employment in a field is primarily in the for-profit corporate sector or in the nonprofit, academic, or government sectors. The higher-paying positions are concentrated in engineering fields in which corporate employers dominate, such as computer, electronics, chemical, and petroleum (though only recently for the last). Conversely, the lower-paying science and engineering fields tend to be those in which employment is concentrated among employers in academic, nonprofit, and government sectors, and most notably the large scientific sector of basic biomedical research.[82]

Third, among those with science and engineering degrees at all levels, early-career remuneration (i.e., within the first five years following receipt of degree) varies a great deal by field.[83] At the top are fields such as computer/mathematical sciences and some fields in engineering. This is especially so for those with recent bachelor's degrees in engineering, because more work for firms than for nonprofits, and because those with engineering bachelor's are normally viewed as fully qualified professionals—unlike bachelor's in scientific fields. Initial median salaries are relatively low in the biosciences (including the biological, agricultural, environmental, and life sciences), where many are employed by nonprofit and governmental organizations with typically lower salary scales.

To give some sense of the wide ranges involved (2008 data), the median annual salary of a recent bachelor's graduate was $59,000 in engineering and $51,000 in computer and mathematical sciences, versus only $30,000 for a recent bachelor's graduate in the biosciences. For recent doctoral degree recipients, the comparable median salaries in 2008 were $86,000 in engineering and $80,000 in computer/mathematical sciences, versus only $50,000 in the biosciences. Oddly enough, it is doctorates in the biosciences that tend to require considerably longer periods of study than those in other fields with higher remuneration, though the lower median salaries may be depressed by the large fractions of recent bioscience PhDs who accept post-doctoral positions with low remuneration.

Fourth—and perhaps most significantly for debates about "shortages"— over the past several decades the trend lines of remuneration for most science and engineering fields have generally been rising more slowly than in other occupations requiring advanced education. Had the opposite been true, this would have represented an important indicator of emerging "shortages" in the science and engineering workforce relative to demand in the labor market.

One insightful analysis by Richard Ellis, published in 2007,[84] examined growth trends in both employment and compensation (the latter in constant dollars to control for inflation) for two time periods: 1995–2002 (a period of booming growth in some science and engineering fields), and in 2003–06, representing the most recent data then available and before the effects of the deep economic recession that began in late 2007. The most striking finding was that in both of these time periods, science and engineering occupations looked quite similar to general trends in the U.S. workforce at *all* levels of skill and occupation. There was, as oft noted, a very large increase in the number of persons employed in some information technology occupations during the first time period from 1995–2002. For example, the number employed as "computer scientists and systems analysts" was an extreme outlier in employment growth during this period, increasing more than 75 percent faster than the number of overall job growth in the overall workforce. Yet the compensation received by this occupational category increased only slightly more rapidly than the overall increase, indicating that growth in the supply of computer scientists and systems largely met the rapid growth in demand.[85] Moreover, compensation in IT occupations has grown far less robustly than in some other occupations, especially in the financial and medical sectors, that require comparable levels of quantitative sophistication. The same is likely the case for other STEM occupations.

The relative attractiveness of non-STEM careers for STEM graduates, described as "diversion" in one report discussed earlier in this chapter, has been an important factor underlying the substantial fractions of science and engineering graduates from leading institutions who have chosen to pursue careers in finance rather than in science and engineering.[86] It remains to be seen whether the financial meltdown that began in 2008 has affected the relative attractiveness of the financial sector for more recent graduates. Lesser growth in numbers but higher growth in compensation also was shown during the early 2000s by "medical scientists," possibly a reflection of the 100 percent increase in the NIH budget between 1998 and 2003, a boom phase that also subsequently disappeared.

Summary: No Consensus, Divergent Interests

In view of ambiguities and confusions described in this chapter, it is understandable that consensus is lacking about the adequacy of the U.S. science and engineering workforce for the country's economic needs. Moreover, it is easy to see how groups with strong self-interests in this subject are able to find apparently credible evidence to support arguments that there are current or looming "shortages" of scientists and engineers, while others can point to substantial data indicating that scientists and engineers are in ample or excess supply in the U.S. labor market.

That is the current sorry state of affairs, but as we saw in chapter 2, it is not a *new* sorry state of affairs. To the contrary, the controversies of today about whether science and engineering labor markets are in shortage or surplus are the latest repetitions of hardy perennials. Similar disputes occurred at different points going back a half century or more. There is much to be learned from the effects of the policies that emerged from these past episodes, from the ultimate accuracy of the differing points of view, and from the roles played by a variety of advocacy and other groups.

CHAPTER 6
The Distinctive U.S. Academic Production Process

Simply producing more engineers and scientists may not be the answer because the labor market for those workers will simply reflect lower wages or, perhaps, greater unemployment for those workers.

—Ben Bernanke, Chairman, Federal Reserve Board[1]

The U.S. academic system is distinctive, even unique, in the ways that it recruits, finances, and produces newly minted scientists and engineers. In most other countries, scientists and engineers receive their education in a limited number of universities, polytechnics, and other postsecondary institutions. There are fewer than two hundred degree-granting institutions in such major advanced countries as the United Kingdom, France, Germany, and Canada.[2] Nearly all of these institutions receive substantial direct financial support for their core educational activities from national, provincial, and sometimes local governments. This means that governments can play an important direct role in deciding upon the numbers of students enrolled, as well as the distribution of these enrollments in different fields of study.

In contrast, the U.S. system of higher education includes well over four thousand institutions, and is far more variegated. As shown in table 6.1, it includes:

- about two hundred research universities, the best of which rank among the finest in the world (see discussion later in this chapter);

• an even larger group of more than seven hundred "doctoral/research" and "master's" universities that award doctoral and master's degrees but have less research activity;

• nearly nine hundred colleges/universities that award primarily four-year baccalaureate degrees, some of which are as outstanding academically as the elite research universities but are devoted primarily to undergraduate education;

• 950 specialized four-year institutions that focus on single fields of study; and

• more than 1,700 two-year institutions such as community and technical colleges.

Table 6.1. Number of Degree-Granting Institutions, by Type

Total	4,474
Research university, very high[a]	99
Research university, high[b]	102
Doctoral/research university[c]	80
Master's[d]	649
Baccalaureate[e]	883
Specialized institutions[e]	950
Two-year	1,711

Source: U.S. Department of Education, National Center for Educational Statistics, *2009 Integrated Postsecondary Education Data System (IPEDS)*, Spring 2010, table 244. Available online at http://nces.ed.gov/programs/digest/d10/tables_3.asp.
   Notes:
   [a] Research universities with a very high level of research activity.
   [b] Research universities with a high level of research activity.
   [c] Institutions that award at least 20 doctor's degrees per year, but did not have a high level of research activity.
   [d] Institutions that award at least 50 master's degrees per year.
   [e] Institutions that primarily emphasize undergraduate education. Also includes institutions classified as four-year under the IPEDS system, which had been classified as two-year in the Carnegie system because they primarily award associate's degrees.
   [f] Special-focus four-year institutions award degrees primarily in single fields of study, such as medicine, business, fine arts, theology, and engineering.

The approximately two hundred U.S. research universities are further divided as to the sources of their core funding. About 140 are "public," meaning that they receive substantial direct financial support from governments, but almost always at the level of the state rather than national government—there are essentially no research universities with direct budgetary support from the U.S. federal government, although all do receive federal grants and contracts in support of research.[3] About sixty research universities are "private"—nonprofit institutions that are financed heavily by student tuition (which can be very high); by income from private endowments (which can be very large); by ongoing contributions from alumni and other supporters; and by government grants and contracts primarily in support of their research activities.

A useful approximation of the set of leading U.S. research universities may be found in the list of sixty research universities, both public and private, that are members of the Association of American Universities (AAU). Inspection of this membership list will show that it includes most of the highly ranked research universities in the world.[4] Of these sixty, some thirty-four are public institutions, while twenty-six are private.[5]

Research universities, both public and private, are responsible for educating the vast majority of scientists and engineers in the United States. The public research universities receive core support from state governments and in turn charge lower tuition fees to students who are state residents. This was a successful model for decades, as demonstrated by the outstanding quality of research-intensive public university systems such as the University of California (now with ten campuses), and by world-class flagship campuses in other states such as Illinois, Washington, Michigan, Texas, Wisconsin, Maryland, Iowa, Georgia, Ohio, Pennsylvania, and elsewhere. It is, however, a model that has been deteriorating noticeably over recent decades, as state governments have curtailed their core support for their state universities. These universities have responded in a variety of ways, including by rapidly increasing the fees charged to their students, by increasing the numbers of out-of-state and international students who pay higher tuition fees, and by encouraging their faculty members to seek increased financial support from federal research grants.

Leading private research universities (e.g., Chicago, Columbia, Cornell, Harvard, Johns Hopkins, MIT, Princeton, Stanford, Yale, and many others) typically charge far higher student fees than their public peers. However, they also provide substantial financial assistance (sometimes known as "discounts"), especially to students from lower-income groups. These subsidies are drawn in part from net revenues from their high tuition rates, from their often very large accumulated endowments, and from ongoing philanthropy. Some of these financial inflows have also been declining in recent years, especially those from endowments.

Yet for both public and private research universities, a large and increasing fraction of their operating revenues actually comes from the national government, though not in the form of core funding as is the case in most other countries. Instead, it is obtained primarily via research grants and contracts from large federal funding agencies that concentrate on science, engineering, and mathematics,[6] especially the National Institutes of Health, National Science Foundation, National Aeronautics and Space Administration (NASA), Department of Energy, and the Department of Defense. Indeed, and again unlike in most other countries, a majority of research supported by the U.S. government is undertaken in research universities rather than in freestanding research institutes.

This funding model represents one of the sources of the U.S. system's strength. It combines high-level research activities with graduate education by directing much of government support for civilian research to universities rather than to freestanding research institutes, as is the practice in many other countries in Europe and Asia. This funding model has been notably successful in engaging graduate students in leading-edge research as part of their PhDs, and has increasingly been emulated in other countries.

The Outstanding Research Performance of U.S. Universities

U.S. research universities are widely admired around the world for their high levels of research productivity. By most measures they remain the leading research universities in the world. One of the most cited international sources, the *Times Higher Education Supplement* (London), ranks the following numbers of U.S. universities among its top-ranked twenty-five institutions by major field as of 2011–12:

- fifteen of the top twenty-five in engineering and technology
- fourteen of the top twenty-five in life sciences
- sixteen of the top twenty-five in the physical sciences[7]

Another widely cited international ranking, known as the Academic Ranking of World Universities (ARWU) produced by the Center for World-Class Universities at Shanghai Jiao Tong University, suggests if anything even greater U.S. predominance:

- twenty of the top twenty-five in engineering/technology and computer science
- nineteen of the top twenty-five in life and agricultural sciences

- eighteen of the top twenty-five in natural science and mathe-matics[8]

These stellar rankings of U.S. research universities depend upon the high quality of their research faculty, but also heavily upon a governmental funding model in which basic research funding is directed toward research universities rather than government research establishments.

Origins and Evolution of the U.S. Research Funding Model

The origins of this model lie in a 1945 report to the president entitled *Science: the Endless Frontier*.[9] The report was prepared in response to a letter from President Roosevelt in November 1944 to Dr. Vannevar Bush, his closest advisor on science and technology issues. The late I. B. Cohen, long-time professor of History of Science at Harvard University, reported in 1994 that a few years before he died Vannevar Bush confided that Bush himself had actually written the 1944 letter signed by Roosevelt that requested this report.[10]

Bush was a distinguished electrical engineer and science administrator, former dean of engineering at MIT and president of the Carnegie Institution of Washington.[11] In 1940 Roosevelt had appointed Bush to be director of the new Office of Scientific Research and Development (OSRD) in the Executive Office of the President,[12] which subsequently played critical roles in the mobilization of U.S. science and engineering talent to meet wartime challenges. Among other achievements, his office initiated and coordinated the Manhattan Project (transferred to the army in 1943), the development of radar and sonar, and the mass production and distribution of penicillin to reduce deaths from wartime injuries.

Vannevar Bush's report in response to the 1944 request signed by President Roosevelt, entitled *Science: The Endless Frontier*, was submitted to President Truman in July 1945, a few months after Roosevelt's death. Its strongest recommendations were that the federal government should begin to provide financial support for both basic scientific research and for student fellowships, and that most of this support should be directed toward the nation's research-oriented universities.

The report noted that basic research at U.S. universities had long depended primarily upon private funding sources, especially university endowments, philanthropy, and foundations, but that these were all showing signs of decline. Meanwhile most students in the sciences were required to finance their own higher education, thereby limiting the numbers able to pursue such studies. Bush observed that the federal government had long provided substantial support for applied science via mission-directed agencies such as the Departments of Agriculture and Commerce, as well as the military and other security-related agencies. He concluded that the time had come for the federal government to begin to provide additional support for basic research, which would contribute not only to scientific understanding but also to the nation's health, prosperity, and security. He also argued that the best place for such research was in the colleges and universities, rather than in government or industrial laboratories:

> These institutions are uniquely qualified by tradition and by their special characteristics to carry on basic research. They are charged with the responsibility of conserving the knowledge accumulated by the past, imparting that knowledge to students, and contributing new knowledge of all kinds. It is chiefly in these institutions that scientists may work in an atmosphere which is relatively free from the adverse pressure of convention, prejudice, or commercial necessity. At their best they provide the scientific worker with a strong sense of solidarity and security, as well as a substantial degree of personal intellectual freedom. All of these factors are of great importance in the development of new knowledge, since much of new knowledge is certain to arouse opposition because of its tendency to challenge current beliefs or practice.

Reflecting the views of much of the scientific community, Bush was at pains to ensure that any such federal support would be free of the government control that had been exercised in wartime (and coordinated by Bush's OSRD). His report argued instead for creation of an independent "National Research Foundation (NRF)." It would include divisions supporting basic research in three areas stretching across a wide range: national defense, medical research, and natural sciences. The NRF would also have a division of scientific personnel and education that would operate the fellowship programs he recommended for undergraduate and graduate study in the sciences.

In the end this organizational recommendation was not adopted by the U.S. government. In particular, the recommended National Research Foundation was never created. Instead, separate support for basic research on national defense issues began to be provided by the new Department of Defense, established in 1947; for medical research via a much-expanded National Institutes of Health (NIH); and for natural sciences research via creation of a new entity, the National Science Foundation (NSF). However, the funding model embodied in the report—federal financial support for both basic research and for student financial support, directed primarily to U.S. universities rather than to government laboratories—formed the core of subsequent NIH and NSF support for research and higher education. Given the global predominance achieved since then by U.S. research universities, it is clear enough that this model of government support for university-based research has paid large dividends.

The Evolution from Joint Production to Joint Funding of Advanced Education

Bush's report emphasized the need for government investment in what it termed a "reasonable number of undergraduate scholarships and graduate fellowships in order to develop scientific talent in American youth." The immediate goal of this kind of support was to reverse the "wartime deficit" of newly educated scientists and engineers that was one result of the universal wartime draft, and who could be "on call in national emergencies" that might arise in the future. He made clear he was not recommending that the government seek to maximize the number of newly minted American scientists: "The plans should be designed to attract into science only that proportion of youthful talent appropriate to the needs of science in relation to the other needs of the nation for high abilities."[13]

The report's support both for basic research and for student scholarships and fellowships at U.S. universities embodied a model of what we might call "joint production" of basic research and advanced education—a "critical linking of support for basic research with the advanced education of aspiring scientists and engineers."[14] At many points, the report emphasized that institutions of higher education and research are the best setting for the conduct of basic research and the appropriate producers of graduates with advanced degrees. For example:

> Publicly and privately supported colleges and universities and the endowed research institutes must furnish both the new scientific knowledge and the trained research workers....
> Government, like industry, is dependent on the colleges, universities, and research institutes to expand the basic scientific frontiers and to furnish trained scientific investigators.[15]

It is important to understand, however, that this model of *joint production* does not mean that Bush was recommending *joint funding* for basic research and graduate education. To the contrary, his report called for two distinct forms of direct federal support: peer-reviewed grants to support basic research at universities; and scholarship or fellowship funding for undergraduate and graduate students that would enable them to learn not only from lectures and laboratory exercises but also through direct experience in working on ongoing research projects led by their faculty. This model is quite different from those in many other advanced industrial countries, where research activities are concentrated in government-financed research laboratories while universities concentrate their efforts upon education.

Hence Bush's recommendation did not contemplate combined grants that would jointly finance both research and graduate education. However the U.S. research funding system as it evolved departed rather substantially from the separate funding model he had envisaged. Initially mechanisms for such separate funding of research and advanced education were established. Indeed, the newly formed National Science Foundation established its Graduate Research Fellowship Program (GRFP) in 1952, only about a year after the creation of the NSF itself. Since then, however, the number of such awards has remained rather small, and indeed followed an odd rollercoaster path: an average of about 500 per year during 1950s, nearly 1,000 per year in the 1960s, back to 500 per year in the 1970s and 1980s, and back up to 1,000 per year in the 1990s and early 2000s. In 2010, the number of GRFP awards was doubled to about 2,000 per year and their stipend levels increased.[16]

The NIH also developed a variety of separate funding programs for advanced education or "training" of future biomedical researchers, now named the Ruth L. Kirschstein National Research Service Awards (NRSA) program. The NRSA program, like the NSF Graduate Fellowships also widely considered to be of very high quality, now provides financial support to about 3,700 individuals each year.[17] The number of these NRSA slots expanded rapidly between 1975 and 1980, but then grew only slowly, increasing less than 20 percent over the thirty-two-year span between 1980 and 2012, while the overall budget of the NIH increased by more than 1,300 percent.[18]

Yet, in spite of the small size and slow growth seen in these NSF and NIH graduate education or "training" programs, the overall number of graduate students and postdocs supported financially by these agencies, and especially by NIH, actually has grown dramatically over this period. How can this be?

The answer to this puzzle is that graduate students and postdocs who receive direct support from these "explicit" fellowship and training programs have come to represent only very small minorities of the total numbers of PhD students and postdocs that actually are supported by the NSF, NIH, and other federal funding agencies. Instead, the bulk of federal support for graduate education evolved into a "joint funding" model, in which support for graduate education comes primarily from *research* grants rather than from the graduate fellowships and scholarships contemplated by Bush's report.

By far the majority of graduate students and postdocs now are supported as "employees" financed by research grant budgets, often termed "research assistants," or RA's. In FY2011, for example, NSF funds supported nearly 44,000 graduate students. Of these only about 14 percent fell under its fellowship/ traineeship programs, while the bulk (i.e., 86 percent of the total) was provided in the form of research assistantships for graduate students employed as research assistants under NSF research and related projects. In that same year NSF funds supported about 7,000 postdocs, of which only about 2 percent were supported by fellowship/traineeship funds.[19]

For the NIH, the best estimate is that the NRSA and other "training" programs support only about 22 percent of the combined total of graduate students and postdocs supported by NIH funds. Once again, the large majority of graduate students and postdocs are supported through the budgets of NIH's far larger research grant programs.[20]

Comparable data are not available for other federal science funders such as the Departments of Defense or Energy, but in all likelihood even larger percentages of the graduate students and postdocs they support receive funding from research grants and contracts rather than from fellowships or training grants.

The conclusion here is not in doubt: the joint production system contemplated by Bush has evolved into a predominantly joint funding system, in which a large majority of the substantial costs of graduate and postdoctoral *education* in science and engineering now is funded under the budgets of federal *research* grants and contracts.

Effects of Joint Funding

This funding structure has significant implications for the way the PhD education system has come to operate. In particular it means that the number of PhD students and postdocs supported by NSF and NIH is determined not by appropriations for fellowships or traineeships, and not by any conscious estimate of educational needs or demand for PhDs in the workforce, but instead overwhelmingly by the size of the *research* budgets of these agencies.

The key point for our purposes is this: increases in *research* grant funding will automatically increase the funds available to support PhD students and postdocs in U.S. research universities. This will occur even if there is no reason to expect that there will be rising demand in the labor market when these growing numbers of PhD scientists and engineers complete their PhDs and postdocs. In effect, the principal driver is the amount of research funding that is available for university-based researchers, with little or no connection to "need" or "demand" for additional researchers with doctoral-level education.

In an important sense the real demand for PhD students and postdoctoral research assistants has evolved to be dominated by demand internal to the universities, primarily as research workers financed by federal research grants. If federal research funding increases, there is no mechanism in place to align the increased numbers of PhD students and postdocs receiving federal support with plausible career opportunities for these PhD scientists.

Basic economic theory tells us that if this non-alignment were to increase the supply of doctoral researchers beyond the demand in the labor market for their services, the result would be increasingly unattractive career prospects for recent PhDs. If the system were "closed," and if accurate information about poor or deteriorating post-PhD career prospects in some science and engineering fields were readily accessible to prospective entering graduate students, some U.S. undergraduates who might otherwise consider pursuing PhDs in such fields would choose to pursue other career options available to them, leading the labor market for doctoral scientists and engineers to gradually "adjust."

However, the system is by no means a closed one, nor is accurate information about career prospects readily accessible—quite to the contrary. First, under federal funding practices university faculty who receive federal research grants are free to recruit globally for PhD students and postdocs to serve as research assistants who can be financed by their grants. Over the past decades an ample recruitment pool has become available internationally, especially from China which has rapidly expanded the number of students completing undergraduate degrees; from India, where graduate programs are less developed; or European countries such as France, Germany, or Italy, in which there are fewer opportunities for postdoctoral research. Moreover, while the stipends provided in the United States for graduate students and postdocs are low relative to other opportunities in the United States, they look quite attractive to applicants from countries in which prevailing wages are much lower (see later discussion).

Over the past two decades, numerous well-informed reports from the National Academies and elsewhere have responsibly urged research grant recipients not to "hire" PhD students and postdocs for whom there is little prospect of research careers.[21] However, the current funding system includes powerful incentives for them to recruit as many research assistants as they can finance from their grants, in the interest of maximizing their prospects for continued federal funding. The 2011 National Research Council report on the NIH training programs described the situation succinctly:

> [I]t also has to be acknowledged that if R01 [research grant] support increases, then the number of trainees will ineluctably increase in lockstep, as happened during the recent doubling of the NIH budget. And if there are insufficient U.S. national trainees, then faculties will aggressively look to international Ph.D.s to fill the gap. No amount of well-intentioned urging of institutions to self-correct will change this equation.[22]

What about information that might cause the system to self-adjust by providing accurate data on recent career experiences? Several studies have confirmed the rather odd fact that there actually is little information available to prospective entering graduate students regarding the career experiences of recent graduates from the programs to which they apply. These studies also have found that the opposite is true of post-baccalaureate programs in other fields such as law, business, and economics, for which the career outcomes of graduates are often publicized as part of recruitment activities.[23]

The most recent such study, undertaken in 2008 by Stephan, examined the websites of fifteen leading graduate programs in each of three large fields of science and engineering—the biomedical sciences, chemistry, and electrical engineering—along with a similar analysis of a fourth field in the social sciences, economics, for comparative purposes. In each case the fifteen selected programs consisted of those ranked 1–10 and 21–25 in the National Research Council's 1995 assessment of research-doctoral programs. Stephan reports that of the forty-five science and engineering graduate program websites examined, only two provided any detailed information about the placements of their recent graduates; another four provided some nonspecific information. (Some seven of the fifteen leading graduate programs in economics provided lists of graduates and their placements.)[24]

In short, the dominant role in supporting science and engineering PhDs that now is played by joint funding from research grants—rather than by fellowship or trainee programs—coupled with the lack of credible career information for entering graduate students, means that there are few effective "signals" from the labor market that would adjust the numbers of "trainees" being supported by federal funds. Instead, other things being equal, the numbers of PhD students and postdocs tend to rise or fall depending on the amount of *research* money available, with little or no reference to whether there is "need" or demand for PhD-trained scientists and engineers beyond the postdoc.

Uncoordinated Intersections with U.S. Visa Policies

The funding structure described previously for support of PhD education and postdoctoral positions intersects in complex and sometimes counterintuitive ways with U.S. policies related to admission of international students and temporary workers. The legislation authorizing fellowship and traineeship support from NSF and NIH generally does require that recipients be U.S. citizens or immigrants with legal permanent residence; those with temporary student visas cannot receive such federal funding. Hence if such fellowships and traineeships represented a large proportion of federal support for PhD students and postdocs, unattractive career prospects in these fields would likely result in fewer applicants. However, similar restrictions do not apply to federal financial support for graduate students and postdocs financed by research grants, which as we have seen now comprise the bulk of federal financial support. This means that the great majority of government-funded "slots" for PhD education and postdocs are entirely open to foreign students and postdocs.

Moreover, federal immigration law and practice has in effect delegated to universities and other educational institutions most of the decision-making about student visas ("F-1" visas). There are no numerical limits on these visas, and while formally speaking they are issued by the State Department, in practice the department will routinely approve an F-1 student visa for nearly any foreign student admitted by a registered U.S. educational institution (with a few exceptions aimed at limiting fraud and security risks).

Postdocs can readily be admitted under a different temporary "Exchange Visitor" visa (known as the "J-1"), which is described as fostering "global understanding through educational and cultural exchanges." But because all exchange visitors "are expected to return to their home country upon completion of their program in order to share their exchange experiences,"[25] these visas are unattractive to those who wish to seek U.S. residence and employment upon completion of their postdocs. Waivers of this return requirement can be obtained, but the waiver process is complicated, expensive, and uncertain. For this and other reasons, the temporary visa for "Specialty Workers" (the "H-1B" visa) has grown in popularity for postdocs, especially after 2001 when the numerical limits that previously applied to these visas were eliminated for universities and nonprofit research institutions (see chapter 4). In short, temporary visas for international students and postdocs are readily available with no numerical limits.

Available data do indeed show that large and increasing fractions of graduate student and postdoctoral "slots" are filled by noncitizens on temporary visas. International student enrollments in science and engineering graduate programs rose from about 110,000 in 1993 to about 158,000 in 2006, and to about 176,000 in 2010.[26] The largest increases have been at the master's level, with smaller increases at the doctoral level. The fields with the largest percentages of international students included engineering and computer science (about 45 percent of total enrollments), physical sciences and mathematics (35–40 percent), and economics (about 50 percent). Enrollments of international graduate students in science and engineering fields are also heavily concentrated as to countries of origin, with only two countries—India and China—accounting for nearly two-thirds, and a few other countries (Taiwan, South Korea, Turkey) dominating the remainder.[27]

The enrollment increases for international students have been on a generally rising trajectory since the 1990s, but not a continuous one; in fact foreign student enrollments declined from 2004 to 2007, likely in response to declining U.S. labor market demand following the high-tech busts that began in 2001, and also possibly in response to perceived increased difficulties in obtaining student and other temporary visas after the 2001 attacks in New York and Washington. These declines were concentrated in computer science and in engineering fields, and proved to be temporary. By 2009, the numbers had rebounded sharply, rising by about one-third since 2007.[28]

Even higher percentages of international participants may now characterize the postdoctoral system that has evolved in the United States. As noted in chapter 5, the empirical data about postdocs are remarkably weak, but all available evidence suggests rapid expansion.

Many postdocs bring new ideas, capabilities, and enthusiasm into labs, as is often noted by supporters of the current system. They are young, energetic, and willing to work very long hours. Yet it must be said that there also are direct economic incentives for employers of postdocs: they continue to be the cheapest source of sophisticated research assistance on federally funded research projects. Federal funding agencies such as NIH and NSF have made genuine efforts to bring postdoc stipends into better balance with other categories, and indeed postdoc stipends are considerably better than in the past. They are, however, still modest relative to postdocs' age and education, and employers often do not provide postdocs with the more costly employment benefits (e.g., retirement plans, and sometimes even health insurance) they make available to other employees, thereby further lowering their cost.

The continuing low remuneration and lengthy terms of postdoctoral positions may be making such appointments less attractive to U.S. students who have completed their PhDs, especially for women as postdoc status has lengthened into the 30–39 age group. Indeed, there is some evidence to suggest that there may be declining interest in the postdoc among those earning PhDs in the United States.[29] In addition, there is concern that the combined length of the PhD and postdoc, which now often extends into the early thirties or later, may be limiting the independent creative contributions that in the past were made by prominent senior biomedical researchers while they were still in their twenties.[30]

In a normal labor market, the declining attractiveness of postdoc positions to U.S. students would be expected to result in fewer applications from recent PhDs, thereby producing a self-correcting negative feedback in the system. However, the structure of the current system (again inadvertently) minimizes any such negative feedback effects.

As noted earlier, universities may use their federal research grants to finance international postdocs (and PhD students), while as also noted earlier current immigration policies provide them with essentially unlimited access to temporary visas for postdocs.[31] (See also discussion in chapter 4.)

This unlimited access to expanding global pools of PhDs likely has the (unintended) effect of limiting any market adjustment mechanisms that would almost certainly have occurred under different circumstances. Why so? As postdoctoral positions have become increasingly unattractive to citizen PhDs, U.S. universities seeking to "hire" them as research workers have not faced any market pressure to improve their attractiveness because they can readily recruit international postdocs. Indeed, many international postdocs are self-recruiting, and some may even bring partial funding from their home governments.

But why would U.S. postdoctoral positions be more attractive to international than to resident PhDs (both citizens and permanent visa-holders)? One reason is that most other countries, including those in Europe, have few postdoctoral opportunities available. A second is that although living stipends for postdocs remain low by U.S. standards for highly educated people, they can at the same time be quite high by the standards of low-income countries such as China or India. Moreover, international PhDs from such countries also likely incur much lower "opportunity costs" than do U.S.-citizen PhDs, that is, the income and benefits they must forego in the course of a postdoc are lower because their prospective earnings at home would be lower, and unlike U.S.-citizen PhDs they do not have the option of seeking regular employment in the United States. Finally, international postdocs benefit from additional incentives that do not accrue to postdocs who are citizens or permanent residents—the possibility of transitioning from a temporary to a permanent visa that allows them if they wish to seek regular U.S. employment, or the prestige of having held an international research position if they decide to return to their home country.

As a result of the intersection between these two uncoordinated federal systems—the first providing government research funding to support postdocs on a global basis, the second providing essentially unlimited numbers of temporary visas for postdocs—the data available suggest that the number of international postdocs has been increasing rapidly. Indeed, they now may constitute a majority of all postdocs in U.S. institutions. These patterns are well-illustrated in figure 6.1, which summarizes the data on postdocs from the NSF's Survey of Graduate Students and Postdoctorates between 1993 and 2006. These data, acknowledged by the NSF to provide only a quite partial count of the true total of postdocs, show that over this span of only thirteen years up to 2006, the number of postdocs increased by nearly 40 percent. Almost all of this increase was accounted for by international postdocs. The largest increases have been in the heavily funded biomedical sciences, and China appears to be the largest source country.

Richard Freeman concisely summarizes the scene for recent PhDs and postdocs in science and engineering as follows:

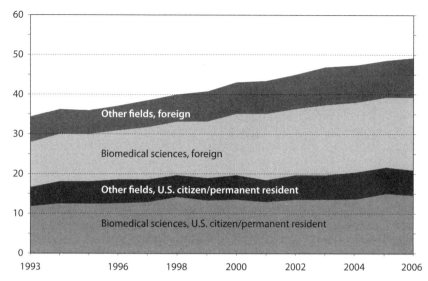

Figure 6.1.
Postdoctoral students at U.S. universities, by field and citizenship status: 1993–2006.
Source: National Science Board, *Science and Engineering Indicators 2010*, ch. 2. Higher Education in Science and Engineering, Postdoctoral Education, fig. 2-24. http://www.nsf.gov/statistics/seind10/c2/c2s4.htm.

The job market has worsened for young workers in S&E fields relative to many other high-level occupations, which discourages US students from going on in S&E, but which still has sufficient rewards to attract large immigrant flows, particularly from developing countries.[32]

Xie and Killewald's analysis of the data leads them to the similar conclusion that "Competition with foreign students steers US students to nonscience fields." They emphasize that it is the academic labor market for PhDs that is particularly problematic. The data show that aggregate academic employment has increased, but they believe this to be misleading because of especially large increases in postdoc and nontenure-track appointments— "positions that lack both independence and security and thus are not truly academic jobs in the conventional sense."[33]

This situation not only had implications for the attractiveness of research careers for U.S.-citizen PhDs, but is also a system in which federal agencies subsidize the advanced education of large numbers of noncitizens who hardly could be "on call in national emergencies"—one of Vannevar Bush's strongest justifications for federal funding of higher education in science and engineering.[34]

Summary and Conclusions

The large and complex U.S. system of higher education rose to dominate basic scientific and engineering research during the decades following World War II. Its predominance was driven in part by heavy federal expenditures on research and development, but a great deal of its success resulted from a research funding model recommended to President Truman in a formative 1945 report by Vannevar Bush. In this approach, federal research funding was directed toward university-based researchers rather than to government research labs, based on peer review of research proposals.

The report also recommended that federal agencies provide support for education and training at these same universities, especially at the graduate level. In effect the report argued for complementary "joint production" of both advanced research and advanced education, supported under separate streams of research and education funding to research universities. It also made clear that education support should be provided only to a "reasonable number," and that the fraction of youthful talent thereby attracted into science should be proportionate to the needs of science in relation to other national needs. Bush's model did not contemplate the later rapid expansion of "joint funding" of research and advanced education, in which financial support for graduate students came to be dominated by funding for research rather than for education and training.

A quartet of elements, presumably unplanned and unintended but well established, subsequently combined to create the current system, in which the numbers of those applying and enrolling in PhD and postdoctoral positions can be sustained and even increased irrespective of whether there is sufficient demand for such highly educated personnel in the labor market. These elements include:

- the dominant use of federal *research* (rather than education or training) funds to support the large majority of graduate students and postdocs;

- the very substantial increases in the magnitudes of such research funding;

- the limited information available to prospective graduate students about career prospects in many PhD fields; and

- easy access to U.S. graduate institutions and federal financial support for rapidly expanding global pools of graduate students and postdocs.

Together these ensure that there are no effective mechanisms by which slack demand for PhDs in the labor market might be expected to affect the numbers who enter into PhD programs and postdocs in these fields.

CHAPTER 7
International Comparisons:
Glass Half-Full, Glass Half-Empty?

As scarce as truth is, the supply has always been in excess of the demand.

—Josh Billings[1]

Over the past two to three decades important shifts have taken place in international comparisons and "rankings" of countries concerning their research and development (R&D) in science and engineering. The same can be said for international comparisons of the effectiveness of science and mathematics education at primary, secondary, and higher levels.

The dominant position of the United States during the decades after World War II is apparent in standard indicators of such comparisons such as the volume of published scientific articles, numbers of patents issued, and production of STEM graduate degrees awarded. As noted earlier, of course, this postwar dominance of the United States itself represented a major shift from the leading positions held by a succession of European countries (Great Britain, France, Germany) over the preceding two centuries.

During the same postwar decades the science and engineering capacities of the Soviet Union expanded rapidly, driven by heavy national investments in military and space research and development, and by production of large numbers of graduates in science, math, and engineering fields. Indeed, as described in earlier chapters, this led to U.S. alarums during the 1950s that the Soviet Union was pulling ahead in these fields, resulting in massive U.S. government investments in the "space race" and in expansion of science and mathematics education. The subsequent economic and political weakening of the USSR during the 1980s, and its ultimate collapse in 1991, sharply reduced science and engineering activities in its successor states of the Russian Federation and other former Soviet republics.

From the 1990s onward, U.S. investments and capabilities in science and engineering continued to expand substantially. Yet the U.S. predominance in these fields that had been achieved during the postwar decades also began to gradually erode, as countries in both Europe and Asia began to catch up.

This narrowing over the past two decades in U.S. dominance in science, engineering, and mathematics has led some commentators to re-sound the alarm that the United States is in danger of falling behind its major competitors for global leadership in these fields. Typically such concerns are linked to other worries about declines (again in relative terms) of U.S. economic performance, as measured for example by sustained massive trade deficits driven by offshore-outsourcing of manufacturing to low-wage countries during a period of rapid economic globalization. Taken together, these trends have led to arguments that the decline (again, we must emphasize, in relative terms) of U.S. dominance in science and engineering portends gloomy economic prospects for the United States, based on arguments that economic competitiveness now is determined by science and technology driving "innovation."[2]

Such views have been expressed most forcefully by some leaders in corporations, higher education, and politics. Indeed, the 2010 follow-up report of the committee that produced the National Academies 2005 report *Rising Above the Gathering Storm* discussed in chapter 1, whose membership was dominated by leaders of major corporations and universities, drew heavily upon international comparisons to reach a provocative and memorable conclusion framed in a ten-word sentence:

"The *Gathering Storm* increasingly appears to be a Category 5."[3]

In support of its conclusion, the 2010 follow-up report offered a list of what it called "factoids" in its Executive Summary:[4]

China is now second in the world in its publication of biomedical research articles, having recently surpassed Japan, the United Kingdom, Germany, Italy, France, Canada and Spain.

In 2009, 51 percent of *United States* patents were awarded to non-United States companies.

Only four of the top ten companies receiving United States patents last year were United States companies.

The World Economic Forum ranks the United States 48th in quality of mathematics and science education.

The legendary Bell Laboratories is now owned by a French company.

In 2000 the number of foreign students studying the physical sciences and engineering in United States graduate schools for the first time surpassed the number of United States students.

In 1998 China produced about 20,000 research articles, but by 2006 the output had reached 83,000 ... overtaking Japan, Germany and the U.K.

Eight of the ten global companies with the largest R&D budgets have established R&D facilities in China, India or both.

In a survey of global firms planning to build new R&D facilities, 77 percent say they will build in China or India.

GE has now located the majority of its R&D personnel outside the United States.

The United States ranks 27th among developed nations in the proportion of college students receiving undergraduate degrees in science or engineering.

The United States ranks 20th in high school completion rate among industrialized nations and 16th in college completion rate.

In less than 15 years, China has moved from 14th place to second place in published research articles (behind the United States).

China's Tsinghua and Peking Universities are the two largest suppliers of students who receive PhD's—in the United States.

The United States has fallen from first to eleventh place in the OECD in the fraction 25–34 year olds that has graduated high school. The older portion of the U.S. workforce ranks first among OECD populations of the same age.

In 2007 China became second only to the United States in the estimated number of people engaged in scientific and engineering research and development.

Such conclusions have been contested as exaggerations by other knowledgeable observers. As discussed in earlier chapters, a number of rounds of similar concerns about the United States falling behind in science and engineering arose in prior decades, and led to responses that contributed to cycles of "alarm/boom/bust." It is quite true that many such claims in the past may have been "crying wolf."

Yet the fact that during these previous cycles the wolf turned out to not be at the door should *not* lead to the conclusion that present concerns also can be predicted to prove unwarranted. There is ample historical evidence that leadership in these fields can shift, sometimes rapidly. As noted earlier, European leadership in science and engineering over an extended period from the latter half of the nineteenth century to the onset of World War II gave way to American leadership in the decades following that war. The relative position of the former Soviet Union also rose rapidly during the postwar period, and then fell behind from the 1990s onward.

No one can see the future, but it would be wise to consider and assess the evidence that the United States is now falling behind its major competitors in science, engineering, and mathematics. Inevitably, available evidence on such a topic is likely to be mixed or ambiguous. How should an objective observer interpret it?

Still the World Leader, but Less Dominant than 2–3 Decades Ago

Any discussion of this topic should emphasize from the outset that the trends described relate only to *relative* decline. The most recent data still show the United States to be the leading center of higher education, basic research, and patenting in the world. The National Science Board provided a balanced appraisal in its most recent report *Science and Engineering Indicators 2012*:

> In most broad aspects of S&T activities, the United States continues to maintain a position of leadership but has experienced a gradual erosion of its position in many specific areas. Two contributing developments are the rapid increase in a broad range of Asian S&T capabilities outside of Japan and the fruition of EU efforts to boost its relative competitiveness in R&D, innovation, and high technology.[5]

R&D expenditures: Consider first the levels of expenditures on research and development (R&D). At a global level such expenditures have been rising more rapidly than economic output. Indeed, through 2009 global R&D expenditures had been on a path that would lead to a doubling in only nine years,[6] although the subsequent global financial crisis no doubt has led to some decline in this rapid rate of increase.

Within these global totals, R&D expenditures in the United States continue to be by far the largest of any country. Of the estimated global R&D

total of $1,276 billion in 2009, the United States alone accounted for nearly $402 billion, or more than 31 percent. The next largest national expenditures on R&D in 2009 were those of China (12 percent), followed by Japan (nearly 11 percent), Germany (6 percent), and France (4 percent). U.S. R&D expenditures exceeded those of the next three largest combined—China, Japan, and Germany. The combined R&D expenditures of all twenty-seven member nations of the European Union accounted for about 23 percent of total R&D expenditures in 2009.[7]

However, since total U.S. GDP also is the highest in the world, it may be argued that comparisons of national R&D expenditures should be assessed not in absolute terms but instead as a percentage of GDP. In this kind of comparison, U.S. R&D expenditures are also high by international comparative standards—nearly 2.9 percent of 2009 GDP—though lower than the percentages in Japan and South Korea, as well as some other smaller countries such as Sweden, Switzerland, Finland, Denmark, and Israel, all of which were 3 percent or higher.[8] In China this ratio of R&D to GDP has been rising rapidly, but from far lower levels—it is up two and a half times in just over a decade, but from a very low base of only 0.6 percent in 1996, rising to 1.7 percent in 2009. Over the past decade China's GDP has also been expanding far more rapidly than that of the United States, averaging on the order of 12 percent per year,[9] meaning that the absolute volume of R&D expenditures in China was rising rapidly. The same general point could be made about trends in South Korea, but not about Japan where the economy has been relatively stagnant since the 1990s.

Several clarifications about the category "Research and Development (R&D)" need to be mentioned here. First, R&D expenditures are usually divided into three types: basic research, applied research, and development. Of these, basic research (part of the "R") is the smallest of the three categories, generally accounting for less than 20 percent. Applied research also accounts for a small fraction—about 20 percent in the U.S. data. It is "development" (the "D" in "R&D") that has long been by far the largest component of R&D, amounting to about 60 percent of the total in the United States over the past decade,[10] and even higher percentages in other countries. To put the matter in another way, only 20 percent of R&D expenditure falls under the rubric of "basic research," while 80 percent is devoted to either "applied research" or "development," of which the latter is by far the largest.

This is relevant to public discourse on the subject, much of which seems, oddly, to focus on trends in the smallest of the three R&D categories, that of basic research. In the U.S. data, this is the only category for which the federal government is the majority funder (at 57 percent) and for which universities are the majority performers (at 56 percent). Meanwhile, businesses performed and funded over 60 percent of the applied research category, and well over 80 percent of "development."[11]

   Another important underlying driver of R&D expenditures is the fact that manufacturing industries almost universally invest more of revenues in R&D (although still only about 4–5 percent, and primarily in "D") than do service sector industries. This implies that countries with large and/or growing manufacturing activities would be expected to show relatively high or increasing expenditures in R&D—the cases of Japan, South Korea, and most recently China offer good examples of economies with very strong or rapidly increasing manufacturing activities and hence high or growing levels of R&D.[12] Meanwhile these R&D-intensive manufacturing activities in the United States have been declining with the offshore outsourcing accompanying the current wave of globalization.

   The patterns of R&D expenditures in most advanced countries are determined primarily by industry rather than by governments. Governments generally are minority direct funders of R&D in their economies. They may provide tax and other incentives in support of R&D, but the bulk of direct R&D funding comes from the business sector. Having said that, it also is the case that many governments, especially in Asia, have adopted policies designed to stimulate and incentivize R&D activities in their societies, driven by the belief that these will be the leaders of their future economic development.[13] Their actions have included both direct and indirect measures:

- effective government control of some key industries, as in the case of China;

- direct subsidies for R&D-intensive manufacturing designed for export;

- indirect but major incentives (financial, capital, and taxation) and sometimes even mandates to ensure that multinational firms invest substantially in R&D activities in return for approval to market their products in these countries;

- investments in government research funding and in rapid expansion of higher education systems.

   The net effects of this combination of rapidly growing GDPs in some Asian countries, coupled with rising ratios of R&D expenditures to GDP, has indeed produced relative shifts in the "shares" of global R&D expenditures. Between 1996 and 2009, the share of global R&D in North America (here combining the United States, Canada, and Mexico) declined from 40 percent to 36 percent, and the EU share from 31 percent to 24 percent. Over the same period, the Asia/Pacific region's share increased from 24 percent to 35 percent.[14] During this period R&D expenditures increased in all of these regions, hence the shifts in global shares reflect differing rates of increase rather than increases in some and declines in others.

Higher Education

What about comparative international trends in higher education? During the latter half of the twentieth century the U.S. share of global higher education at all levels was very high. The most comprehensive category of higher education, known as "tertiary" education, is defined by international convention to include all higher education at or above the level of U.S. technical schools or associate degrees. By this measure, the United States in 1970 produced more than 30 percent of all "tertiary" education graduates in the world.[15] If we narrow our lens to examine only the highest level of "doctoral" education, and only in science and engineering fields, U.S. universities in that year accounted for more than 50 percent of such doctorates.[16]

This U.S. super-dominance of global higher education was driven by a number of specifically twentieth-century elements:

- The development of mass higher education at all levels in the United States during the 1950s and 1960s, and its subsequent expansion;

- The wartime damage to universities and scientific research in Europe, which previously had dominated higher education in science and engineering, along with movement to the United States of many leading European scientists escaping the depredations of the Nazis and the war and subsequent Cold War in Europe;

- Massive expansion of U.S. government funding of research and doctoral education in science and mathematics in response to the shock of Sputnik 1 (see chapter 2);

- Decades of political instability, slow economic growth, and weak university-level education and research in post-revolutionary China.

Within the past two to three decades these drivers of American postwar exceptionalism in higher education have begun to wane, driven by the recovery of economic prosperity and of science and technology in the war- and revolution-battered countries of Europe and Asia. Some of these recoveries of course were actively supported by U.S. government policies such as the Marshall Plan, the Allied occupation of Japan, and promotion of trade globalization. In addition, the same period has seen the emergence of dynamic economic growth and educational expansion in other countries, especially in Asia, including South Korea, Taiwan, China, and most recently India.

In the latter two cases of China and India, the huge scale of their populations (each more than four times the size of the 2013 U.S. population of over 315 million) means that even small increases in the percentages with science and engineering education can represent very large absolute and relative numbers. The most dramatic growth in this period has been in China, in which the government decided to promote extraordinary expansion of higher education and invested heavily in doing so. Between 1980 and 2000, China's fraction of the world's tertiary-educated population increased from about 5 percent to more than 12 percent, and India's, from 4 percent to 8 percent.[17]

Yet even with notable narrowing of the very large gaps that prevailed only a few decades ago, the United States apparently continues to have the largest fraction of the population with tertiary education. In 2009 the U.S. accounted for some 25.8 percent of the estimated 222 million with tertiary education in the G-20 countries.[18] This figure was well over twice those for the countries with the next largest percentages—12.1 percent in China (for 2000, the latest available, and hence no doubt considerably larger now), and 11.4 percent in Japan—and five times larger than all other reporting countries, none of which report more than 5 percent.[19]

First University Degrees

As noted earlier, the United States led development of mass higher education during the 1950s and 1960s, but other countries began to catch up in subsequent decades by rapidly expanding the percentage of their young adult cohorts obtaining first university (or "bachelor's") degrees. In 1975, the United States exceeded all other countries in this indicator, with 4.72 bachelor's degrees per hundred population aged 20–24 years. In the years that followed, the U.S. ratio continued to increase, but the rates of increase were much higher in a number of countries (mostly in Europe, with a few in Asia such as Taiwan and South Korea) that by 2005 had overtaken the United States in the ratio of bachelor's degrees to the population aged 20–24.

Meanwhile, China began the massive expansion of its higher education system (to be sure with many unanswered questions as to quality). In 1990 there were only 0.21 bachelor's degrees per hundred 20–24-year-olds in China; by 2005 this proportion had increased nearly sevenfold, to 1.45. Obviously this was still a small ratio when compared to the 6.83 in the United States for that same year (not to mention the 11.30 figure for Taiwan). Yet the size of the Chinese cohort aged 20–24 was so enormous that the absolute number of bachelor's degrees in China began to converge with the numbers awarded in the United States, given its much smaller population.

Chinese higher education is also distinctive in another way: its very heavy concentration upon science and engineering, as discussed in greater detail in the section immediately following. The arithmetic of this concentration

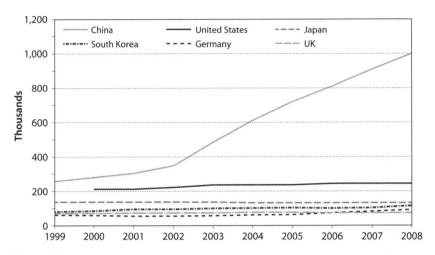

Figure 7.1.
First university natural sciences and engineering degrees, by selected countries,
1999–2008.
Source: National Science Board, *Science and Engineering Indicators 2012*, p. 2-33, fig. 2-27.

upon natural sciences and engineering, combined with very rapid expansion in undergraduate education in general, produced dramatic growth in the number of Chinese first degrees in these fields, as shown in figure 7.1.

Differing Proportions in Natural Sciences and Engineering

For our purposes here it is also important to understand that even as it led the development of mass university education, bachelor's degrees awarded by U.S. universities have long included a smaller fraction of concentrators (or "majors") in the fields of natural sciences and engineering (NS&E) than have those awarded by universities in many other countries. In 1975, for example, about 16 percent of U.S. degrees were in this NS&E category—a level comparable to Austria, Denmark, and Belgium, but well below those in most other advanced industrial countries for which such data were available. For example, the percentage of first university degrees in the NS&E category for that same year in other countries with large populations and economies were: 46 percent in France, 38 percent in South Korea, 35 percent in Taiwan, 32 percent in Germany, and 41 percent in the United Kingdom.[20]

As higher education in general expanded rapidly in these countries, their higher percentages of NS&E graduates tended to decline somewhat but still remained higher than the U.S. ratio, which was (remarkably) stable at about 16 percent.[21] The arithmetically inevitable result was higher ratios of NS&E degrees per hundred persons aged 20–24 in most developed countries than in the United States. Indeed, research by Burelli and Rapoport at the National Science Foundation indicates that the dominant reason for the increases in this measure was not a shift toward higher percentages of bachelor's degrees in NS&E fields, but rather simply the expansion in the overall numbers of bachelor's degrees awarded.[22]

The most recent comparative international data available, for about the year 2006, indicate that for the United States the ratio of NS&E degrees was 15.6 percent, while the ratios continued to be much higher for countries such as France (26.8 percent), South Korea (36.5 percent), Taiwan (37.2 percent), Germany (29.1 percent), and the United Kingdom (22.4 percent).[23]

The same source reports that the percentage of NS&E degrees in China was remarkably high: 44.4 percent.[24] As will be discussed in greater detail, this extraordinarily high total was due primarily to an astounding 31.2 percent of all bachelor's degrees in China reported to be in engineering. Only Singapore, a city-state of less than 275 square miles or 700 square kilometers, with only 5 million inhabitants, reported a slightly higher percentage of bachelor's degrees in engineering, at 33.9%. In comparison, the percentage of bachelor's degrees in the natural sciences in China was far lower than that for engineering—only 13.2 percent—and not markedly higher than that of the United States at 11.2 percent.

There is a considerable debate in the analytic literature about Chinese data on degrees in "engineering." Several research groups have produced analyses raising numerous questions, including whether the large Chinese numbers reflect comparable levels of higher education, include the same fields of study, and reflect comparable standards of quality as those of the countries to which they are compared.[25] For example, it appears that the Chinese data on "engineering" degrees may include two-year technical training certifications equivalent to associate degrees in the United States, and the term "engineer" may be interpreted in China to include what might be considered "technician" or "mechanic" in other countries.

Appropriate adjustments should certainly be applied before international comparisons are made. However, it still seems likely that an exceptionally high percentage of legitimate "bachelor's" degrees in China are in engineering fields. In part this is driven by government decisions that affect the allocation of specialization fields in Chinese higher education. It is often remarked that since the beginning of the Chinese Communist revolution the political leadership has included large numbers with engineering backgrounds.

In addition, the massive expansion of infrastructure under way in China has created enormous labor market demand for engineering skills. Consider

Table 7.1. Growth of Infrastructure between 1997 and 2007,
United States and China

Length in Miles	United States	China
Interstate/Expressway	608	30,519
Navigable Channels	(680)	8,510
Rail	(4,030)	7,436

Source: Hal Salzman and Leonard Lynn, "Engineering and Engineering Skills:
What's *really* needed for global competitiveness," Paper presented at Annual Meet-
ings of the Association for Public Policy Analysis and Management, November 4,
2010, Boston, MA, p. 7.
   Data Sources: For United States: Bureau of Transport Statistics, U.S. Department
of Transportation, *National Transport Statistics*, 2009. For China: National Bureau
of Statistics of China, *China Statistical Yearbook*, 2008.

for example the magnitudes of construction of high-speed roads, railways,
and navigable channels that was undertaken in China between 1997 and
2007, compared with comparable figures for the United States, as summarized in table 7.1. During this eleven-year period, over 30,000 miles of new
high-speed highways were built in China, versus just over 600 in the United
States. Over the same period nearly 7,500 miles of railways were constructed
in China versus a decline of over 4,000 miles in the United States. Infrastructure projects of this type are heavy employers of engineering talent,
primarily in civil engineering, hence the massive expansion of Chinese infrastructure has produced an enormous domestic demand for engineers to
plan, design, and supervise their construction. In the United States in contrast there was, if anything, dis-investment in infrastructure, especially in the
rail sector.

International Differences in Student Choice of Concentrations

Meanwhile, several unusual attributes of U.S. higher education also warrant
some special attention. As noted in chapter 6 of this volume, the U.S. system
of higher education is a large, highly diversified, and atypical amalgam of
literally thousands of institutions. These include world-class research universities with strong PhD programs; professional schools in law and medicine; "comprehensive" and "masters-focused" universities combining both
undergraduate and graduate degree programs; undergraduate "liberal arts"
colleges; and community colleges concentrating upon the first two years
of postsecondary education. There is also a large and rapidly growing higher
education sector that is operated by for-profit corporations.

Without digressing into a lengthy discussion of its strengths and weaknesses, two unusual aspects of the U.S. higher education system are worth recalling in any discussion of student choice of concentration or "major." First, unlike most governments, the U.S. government is able to exercise rather little influence over the percentages of students in the United States who choose to pursue higher education degrees in Natural Sciences and Engineering (NS&E)—those decisions are made by the students themselves and by their educational institutions. The available survey evidence suggests (see figure 7.2) that about one-third of entering freshmen have for many years been expressing the intention of majoring in a STEM field (the NS&E fields plus the social/behavioral sciences combined). This has risen to nearly 40 percent in the most recent data for 2012. If we back out the 11 percent in this group who intend to major in the social/behavior sciences, nearly 29 percent of entering freshmen in 2012 indicated an intention to major in an NS&E field.[26]

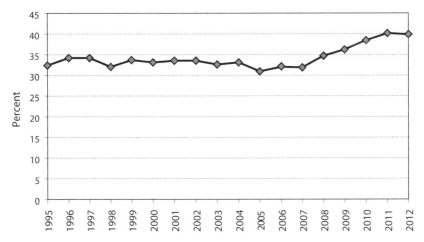

Figure 7.2.

Percentage entering freshmen intending STEM major, 1995–2012

Source: Historical series calculated and compiled by author from: for data from 1995 to 2010, National Science Board, Science and Engineering Indicators 2012, appendix table 2-12, available online at http://www.nsf.gov/statistics/seind12/appendix.htm.

For 2011 data, J. H. Pryor, K. Eagan, Blake L. Palucki, S. Hurtado, J. Berdan, and M. H. Case. *The American Freshman: National Norms Fall 2012 Expanded Tables* (Los Angeles: Higher Education Research Institute, UCLA, 2013), pp. 26–27.

For 2012 data, J. H. Pryor, L. DeAngelo, Blake L. Palucki, S. Hurtado, and S. Tran. *The American Freshman: National Norms for Fall 2011 Expanded Tables* (Los Angeles: Higher Education Research Institute, UCLA, 2012), pp. 37–38. Available online at http://heri.ucla.edu/publications-main.php.

Intentions may not determine outcomes, of course, and the second unusual characteristic of U.S. undergraduate education is the four-year "liberal arts" format that prevails for the first degree, which means that most students begin their postsecondary studies without firm commitments as to their ultimate field of concentration or "major." Typically they are not required to make such a choice until the end of the second year of their four-year degree, although engineering is an exception in which most must declare their choice during the first year and sometimes even at the time of application for admission. This liberal arts tradition allows for much re-assortment of students among concentrations or majors during the first two years of the four-year baccalaureate degree.

To examine how much re-assortment actually takes place we must of course look at longitudinal data for student cohorts who became freshmen at least five to six years earlier so that their decisions about majors can be tracked to graduation. Two such analyses, one by the National Science Board[27] (NSB) and the other by Hal Salzman,[28] use somewhat different strategies to follow initial and ultimate choices of majors by new freshmen in 2004. They both find considerable "fluidity" in the paths followed by U.S. freshmen, and far more than in most other countries.

The NSB data show that overall about 236,000 of 1,388,000 entering freshmen in 2004 had selected an NS&E field[29] as their major (about 17 percent of the total), but by 2009 about 80,000 of these 236,000 (about one-third overall, but with varying percentages across NS&E fields) had shifted out to non-NS&E majors. Much larger numbers of entering freshmen reported initial choices of non-NS&E majors (652,000, or about half) or were "missing/undeclared" (392,000 or 28 percent). Of these, more than 100,000 had shifted into NS&E fields by 2009, but most had selected majors outside of NS&E fields. This re-sorting of freshmen majors resulted in a net inflow into NS&E majors on the order of 20,000–25,000[30] between 2004 and 2009.

Salzman's more detailed analyses, which use a slightly different definition of "science and engineering,"[31] found that about 69,000 entering freshmen in 2004 had left the NS&E majors that they initially declared, while about 81,000 had entered these majors and graduated—again suggesting a small net increase, in this case of about 12,000.[32] Salzman concludes that this fluid process of choice of specialization or major in U.S. undergraduate education can best be characterized by the metaphor of "pathways" rather than by the popular usage of the more unidirectional metaphor of a "pipeline," or more often a "leaking pipeline." He also suggests that it may actually be this "loose coupling between S&E disciplines and S&E careers that provides the U.S. some of its dynamism, innovativeness, and creativity."[33]

This U.S. structure is quite different from bachelor's degrees in most other countries, both developed and developing, which provide a three-year degree characterized by strong specialization from the very beginning and limited opportunities for students to shift to other fields. Moreover, in many such countries the university system is financed largely by the government, and government agencies have considerable influence on the distribution of specializations pursued by university students.

Comparable population data are not yet available from the census rounds of 2010/2011, but other evidence suggests that the absolute number of bachelor's degrees awarded in natural sciences and engineering have been relatively flat in most countries during the past decade, but has expanded greatly in China (see figure 7.1). According to one of the more authoritative sources, of the more than 3 million first university degrees awarded globally in NS&E around 2008 (an incomplete count including only locations for which fairly recent data are available), China's percentage had expanded very rapidly to account for nearly one-third of the world total, representing just over 1 million such degrees.[34]

It is however easy to misinterpret such information. The real story about first university degrees in natural science and engineering in countries such as Singapore, China, and South Korea actually is about *engineering*. While all of these countries do report very high percentages of first degrees in the natural sciences and engineering (about 46 percent, 44 percent, and 37 percent respectively), these high percentages are driven mainly by their exceptionally high percentages of degrees in engineering rather than natural science. Indeed, as noted earlier engineering degrees alone comprise about 34 percent, 31 percent, and 25 percent of all first university degrees in Singapore, China, and South Korea respectively.[35]

The Special Case of the PhD

As previously noted, in 1970 about half of the world's PhDs in science and engineering were awarded by U.S. universities. This percentage remained very high until the turn of the twenty-first century, when China began a concerted initiative to rapidly expand the number of Chinese universities and the production of PhD degrees awarded in China. The results can be seen clearly in figure 7.3.

Over this same period, the number of NS&E doctoral degrees in the United States, which had risen slightly in the 1990s, first declined slightly in the early years of the twenty-first century and then rose substantially up to 2008—the latter increase attributable in large part to increased numbers of U.S. PhDs awarded to foreign students, primarily those from several Asian countries including China that produce high percentages of their undergraduate degrees in these fields. There also were increases in PhD production

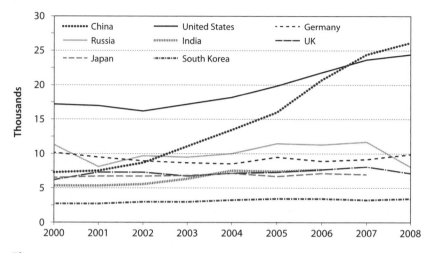

Figure 7.3.
Natural sciences and engineering doctoral degrees, by selected country: 2000–2008.
Source: National Science Board, *Science and Engineering Indicators 2012*, p. 2-35, fig. 2-28.

in countries such as Japan, UK, South Korea, and India, though not in Germany.

Still, the obvious outlier in this graph is China, where the number of NS&E PhD degrees awarded rose rapidly during the 1990s, accelerated to a steeply rising curve beginning around 2001, and grew larger than the still-rising U.S. number somewhere around 2007. As is also the case for engineering degrees, there is considerable uncertainty as to the comparability of Chinese PhDs in terms of intensity and quality. Unlike the stellar performance of U.S. universities in the internationally comparative rankings of universities (see discussion in chapter 6), only a few Chinese universities yet rank among the world's top fifty or even the top one hundred. The rankings produced by *The Times* (London) Higher Education Supplement for 2011–2012 include no Chinese universities among the leading fifty in the physical sciences and in the life sciences, and only four among the top fifty in engineering (of which two are in Hong Kong).[36] The rankings by Shanghai Jiao Tong University[37] show only one Chinese university (Peking) among the top one hundred in the natural sciences and mathematics. The same is true for engineering, and no Chinese universities appear in the top one hundred in this source's rankings in the life sciences.

Nonetheless, the heavy funding and rapid expansion of Chinese higher education suggest that over time more Chinese universities will begin to be ranked among global leaders. The rapid expansion of Chinese higher education also has produced insufficiencies in the number of high-quality faculty members. National and provincial governments in China are energetically recruiting additional faculty from abroad, offering attractive inducements to nationals of other countries as well as to Chinese nationals who earned their higher degrees abroad and chose to stay on in those countries. Over time such recruitment should lead to rising quality standards of PhD programs, and if the recent remarkable expansion of Chinese undergraduate education continues, the rapidly growing numbers of recent PhDs from Chinese universities are likely to experience attractive career opportunities.

Previously, in the 1990s, and of course at a much smaller scale, the government of Japan also decided to sharply increase the number of NS&E PhDs and postdocs in Japanese universities. It appears however that the policy was not carefully related to demand for PhDs in Japan. According to an article in *Nature* magazine in 2011:

> Of all the countries in which to graduate with a science PhD, Japan is arguably one of the worst. In the 1990s, the government set a policy to triple the number of postdocs to 10,000, and stepped up PhD recruitment to meet that goal. The policy was meant to bring Japan's science capacity up to match that of the West—but is now much criticized because, although it quickly succeeded, it gave little thought to where all those postdocs were going to end up....
>
> Academia didn't want them: the number of 18-year-olds entering higher education has been dropping, so universities don't need the staff. Neither does Japanese industry, which has traditionally preferred young, fresh bachelor's graduates who can be trained on the job. The science and education ministry couldn't even sell them off when, in 2009, it started offering companies around Y4 million (US$47,000) each to take on some of the country's 18,000 unemployed postdoctoral students ...
>
> This means there are few jobs for the current crop of PhDs.[38]

Drivers of Trends

Much concern has been expressed that the postwar U.S. domination of global R&D and higher education in science and engineering has been waning. The quantitative evidence available does indeed suggest that growth in these areas has been more rapid in Europe and in Asia than in the United States over the past two decades. One way to think about this is that as Europe prospered under the umbrella of the European Community and later European Union, both higher education and R&D began to recover from the many disasters that afflicted them during the first half of the twentieth century. Over the same period, and particularly beginning during the 1990s, very large Asian countries such as China and India have broken the bonds of ideology or government economic control that previously had led to decades of slow economic growth and even stagnation. In some cases, especially in China over the past decade, there have been truly spectacular rates of expansion in higher education, albeit admittedly from a very low base. It remains to be seen whether these degrees will prove to be of high quality, and whether those who have earned them will find attractive careers paths.

Summary: The Glass Half-Full, or Half-Empty?

The United States continues to be the world leader both in R&D and in higher education in science and engineering. However, the globally predominant roles it held in these spheres for several decades following World War II, challenged only by the former Soviet Union in some fields, have begun to wane as other prosperous and rapidly growing countries have begun to catch up. This is especially true for China, although rapid expansion in that country is taking place from a very low base. The long-term consequences of this shift for comparative levels of innovation, economic growth, and political influence remain to be seen, and there is much concern about some of the symptoms of malaise in the current U.S. system.

In the final chapter, we turn to discussion of how the U.S. system supporting science and engineering research and education, so successful for so many decades, may be in the process of veering badly off track, and offer some suggestions of mid-course corrections that might be worthy of consideration.

CHAPTER 8
Making Things Work Better

If we don't change the direction in which we are headed,
we will end up where we are going.

　　—Chinese proverb

There is a compelling case for public support of science and engineering. This is especially so for basic scientific research, which is not only critical to the national interest and future well-being, but also produces what are called "public goods" that are unlikely to receive much support from non-governmental sources.

A *public good* such as scientific knowledge has two key properties that are obvious once stated: First, its consumption by one person does not reduce its availability for consumption by any other person (economists call this characteristic "non-rivalrous"). Second, scientific knowledge cannot easily be restricted to those who pay for it (economists call this "non-excludability")—partly because openness is central to scientific values and partly because scientists gain personal rewards of professional prestige from openly sharing their work with other scientists. Indeed, it has been convincingly argued that knowledge, including scientific knowledge, is not only a public good, but even a "*global* public good" that benefits the entire world.[1]

These characteristics mean, however, that market mechanisms can be expected to under-invest in public goods such as basic scientific research. Corporations cannot justify large investments in public goods because they cannot capture their economic benefits and hence cannot expect a reasonable investment return. In addition, basic scientific research is a very long-term proposition, and any economic returns it produces are therefore likely to be well beyond the time horizon of even the most longsighted corporate leaders.

This may be contrasted to some degree with "development" activities (the "D" of "R&D") for which engineering is so central. Here much of any economic value that emerges can indeed be captured by those paying for it, in the form of "intellectual property" such as patents and proprietary knowledge that can be excluded from use by competitors. This contrast largely explains the preference of corporations to invest primarily in development and applied research, but much less so in basic research.

The "public good" nature of basic research makes it a natural and appropriate recipient of generous public support. And indeed, as we have seen, public support for U.S. science and engineering since World War II has been generous indeed.

Moreover, this support has by any standards been a roaring success story. Over this period American research and development activities grew rapidly in quantity, and more importantly in quality—so much so that by nearly every measure the United States became the dominant global leader. Still, as in many other success stories, the system that undergirds U.S. research and education in science and engineering has over time also accumulated some unplanned structural elements and unintended negative consequences that are now posing real challenges to its continued success.

In this chapter we address the strongly positive aspects of this system, discussing first the evolution of U.S. success in science and engineering since World War II. We then turn to those structural elements of the system that have been producing progressively more problematic outcomes, while at the same time rendering the remarkable scientific productivity of the system increasingly vulnerable to potentially destructive instabilities.

This leads to a closing discussion of whether it is feasible to formulate cautious and incremental adjustments to the current system that would both maximize the positive and minimize the negative. This discussion would be incomplete if we do not consider the risks inherent to such adjustments for the successful U.S. system of basic research. It also requires a reality check—might the current structure, whatever its problems, be so entrenched as to be immutable or at least powerfully resistant to such incremental changes?

The Indisputable Success of U.S. Science and Engineering

On balance, it is fair to say that the U.S. system has been a triumph of major proportions. A series of decisions at the end of World War II to provide public funding for both research and higher education in science and engineering productively supported the individual and collective creativity and energy of U.S. scientists and engineers. Beginning in the 1950s and continuing to the present, very substantial federal funds have been provided that have enabled the United States to become the leading contributor to these fields.

A triumph, yes, but so as to minimize undue triumphalism it is only fair to acknowledge that U.S. dominance during this extended period was also due in part to the tribulations that impeded postwar science and engineering in other developed countries that had been previous leaders in these fields, including Germany, France, and the United Kingdom. Other potential contenders for global leadership in science and engineering such as Japan also faced serious limitations resulting from World War II. Even nations that have more recently become strong in some fields of advanced science and engineering, such as South Korea, Taiwan, Singapore, and China, also had been limited for decades after World War II by constraints that were both economic and political in character.

Put simply, as economic prosperity and dynamism returned to Western Europe and Japan from the 1960s on, these countries began to catch up in science and engineering. More recently the East Asian countries mentioned earlier have followed suit, as they became less limited by widespread poverty and political instabilities. Nonetheless, by most measures U.S. science and engineering continues to be the strongest in the world, although it would be unrealistic to expect that this will remain so indefinitely.

With the 20/20 vision of hindsight, we can say that the mechanisms and large flows of U.S. government expenditures on science and engineering fields since 1950 have paid very handsome dividends. These expenditures were of course essential, but success was not simply a result of large infusions of money, though large indeed they were. Fundamentally the advance of research in science and engineering is driven by the hard work, commitment, creativity, indeed the brilliance of those who pursue it, and in this respect U.S. scientists and engineers have been exceptional.

Other developments also played important roles. In particular, a number of critical circumstances and policy decisions laid the framework for rapid U.S. advances in scientific and engineering fields:

- the massive postwar expansion in U.S. higher education;

- a postwar political consensus that science and engineering represented major contributors to national security during the Cold War;

- the political panic that led to massive support for science and engineering following the launch of Sputnik 1;

- an academic culture that values creative thinking and rewards achievement rather than political or social connections (by no means true everywhere);

- the operation of strong and creative corporate research laboratories such as those of Bell Labs, IBM Research, and several large pharmaceutical companies, in some cases financed by the negotiated rate base of regulated monopolies such as AT&T; and

- the development of the venture capital sector and other flexible elements of the U.S. financial system that played critical roles in the creation of Silicon Valley and the biotech industry.

Nonetheless, federal support for science and engineering did indeed play a major role. Consider that over this relatively short period of four to five decades from 1950:

- The National Science Foundation evolved from a small, fragile new entity to a science and engineering powerhouse, with annual budgets that now top $7 billion. Over the same period, the National Institutes of Health evolved from a relatively small intramural research institute to a giant funder of basic biomedical research, with an annual budget now exceeding $30 billion. Together these federal funders of basic science have supported research by tens of thousands of researchers in many hundreds of universities and research institutes.

- The Department of Defense became a heavy funder of both basic and applied research and development, as the potentials of science and technology for national security became evident and as the U.S. military increasingly came to depend upon military superiority via technology rather than sheer size. Indeed, the Internet itself, now a major factor in the global spread of science and engineering, had its origins in the Department of Defense.

- The 1958 creation of NASA, admittedly born of what in retrospect seems an excessive political reaction to Sputnik as discussed in chapter 2, established a federal agency that ultimately demonstrated a stunning capacity to accomplish previously impossible technological feats such as the moon landing.

- The Department of Energy, created in 1977 as an amalgam of existing entities,[2] supported major research and development programs via the network of national laboratories under its umbrella, and later played a formative role in an initiative that culminated in the Human Genome Project.[3]

The Role of Policy as a Wellspring of Success

The past and current success of this system of federal support for research and development owes much to the recommendations put forward in Vannevar Bush's 1945 report to President Truman, *Science: the Endless Frontier*, discussed in greater detail in chapter 6. In an important sense this 1945 report has been the "foundational text" of the current system.

One of the most important of the structural decisions taken in the 1950s was to allocate the bulk of the federal government's basic research funds to work conducted in universities and independent institutes, selected by peer review, rather than primarily in government laboratories as was the practice in some other countries such as France and Germany. This was one of the "Five Fundamentals" that Vannevar Bush argued would be critical to the success of government support of civilian research. Since these Five Fundamental principles are often invoked—but sometimes not accurately—it is appropriate to quote them here directly from Bush's report:

1.   Whatever the extent of support may be, there must be stability of funds over a period of years so that long-range programs may be undertaken.

2.   The agency to administer such funds should be composed of citizens selected only on the basis of their interest in and capacity to promote the work of the agency. They should be persons of broad interest in and understanding of the peculiarities of scientific research and education.

3.   The agency should promote research through contracts or grants to organizations outside the Federal Government. It should not operate any laboratories of its own.

4.   Support of basic research in the public and private colleges, universities, and research institutes must leave the internal control of policy, personnel, and the method and scope of the research to the institutions themselves. This is of the utmost importance.

5.   While assuring complete independence and freedom for the nature, scope, and methodology of research carried on in the institutions receiving public funds, and while retaining discretion in the allocation of funds among such institutions, the Foundation proposed herein must be responsible to the President and the Congress. Only through such responsibility can we maintain the proper relationship between science and other aspects of a democratic system. The usual controls of audits, reports, budgeting, and the like, should, of course, apply to the administrative and fiscal operations

of the Foundation, subject, however, to such adjustments in proce-
dure as are necessary to meet the special requirements of research.[4]

Over the ensuing six to seven decades, the U.S. government has fulfilled
many, but by no means all, of these recommendations. In particular it has
implemented major elements of four of the five fundamentals emphasized
by Bush, specifically those numbered (2)–(5) above, albeit in somewhat different forms than those he recommended:

Re: Fundamental Principle #2: The National Science Foundation
and the National Institutes of Health have indeed been led by leading scientific experts and staffed by scientists and engineers with
real understanding of the fields involved. Moreover, both have depended heavily upon external expert reviews ("peer review") in the
selection of grantees.

Re: Fundamental Principle #3: federal civilian research has indeed
been funded via grants and contracts to organizations outside the
federal government, although the Department of Energy manages
a large network of "national laboratories," and with a small part of
its budget NIH operates a substantial and high-quality "intramural"
research program with a large number of its own laboratories.

Re: Fundamental Principle #4: Universities and research institutes
that have received federal research funding have indeed controlled
"policy, personnel, and the method and scope of the research."

Re: Fundamental Principle #5: Institutions receiving public research
funds have retained "independence and freedom," while the funding
agencies have remained responsible to the president and Congress.

However, it also is important to recognize that the United States also has
failed rather badly in implementing Bush's Fundamental Principle #1, that
of providing "stability of funds over a period of years." This has led to significant negative impacts upon the quality and attractiveness of the fields
involved. The successive cycles of alarm/boom/bust reviewed in this book
demonstrate both the failure and the negative consequences that result.

Current Symptoms of Malaise

Notwithstanding its indisputable success since 1950, U.S. science and engineering has more recently been showing numerous symptoms of malaise in
the system discussed in earlier chapters. In brief summary:

Structured for Continuous Expansion That Appears to Be Unsustainable

While U.S. science and engineering research continues to be heavily funded by the federal government, especially so in the biomedical sciences,[5] the system appears to have a tendency to expand beyond whatever funds are available—no matter how large. We have seen that when the 14–15 percent annual budget increases of the NIH "doubling" period ended in 2003, renewed alarms were sounded about a new "funding crisis," notwithstanding the fact that available funds then were twice as large as those of only five years earlier (somewhat less if adjusted for inflation). One well-informed analysis of the biomedical research system, which is by far the most heavily funded area of U.S. science, describes it as having evolved a structure that is unstable without sustained growth of at least 6 percent in the NIH appropriation—a growth level that increasingly appears questionable in view of the looming constraints on overall federal budgets.[6] Another well-informed participant in the biomedical research system diagnoses the malaise symptoms as attributable to an "addiction to rampant expansionism."[7]

Instabilities of Research Funding and Careers

The propensity of the U.S. political system to produce booms and busts in federal funding for science has been discussed in some detail. When combined with the structure of the research system, this tends to generate far more research proposals than can be funded. This is especially evident after the waning of a funding boom, as the number of grant applicants rises due to the increased numbers of PhDs and postdocs who were earlier supported as by-products of the research funding boom. The probability of obtaining funding for a given grant proposal then declines, leading individual researchers and institutions to submit increasing numbers of proposals to reduce the risks of gaps in funding that threaten the research careers and laboratories of even accomplished senior investigators. There also is concern that low success rates lead researchers to "play it safe" by submitting more conservative proposals that are more likely to be funded but less likely to advance the frontiers of science.[8] Finally, the peer review system requires increasing time and effort from researchers to review the expanded number of proposals being submitted.

Lengthening Advanced Training and Unattractive Career Paths

The number of years involved in PhD and postdoctoral research training has grown over the past decades, and the requirement for subsequent postdoctoral

196 • Chapter 8

training also has been expanding. In some fields, initial career positions after the postdoc are increasingly difficult to obtain, as are research grants for junior researchers in academic institutions. Indeed, increasingly, "young" academic researchers are actually approaching middle-age before they can obtain their first career positions and independent research support, leading to concern about the loss of the major scientific contributions by young scientists in previous generations. Relative to other careers that require substantial education after the baccalaureate, career paths for junior scientists in many fields appear to have become unattractive.[9] Oddly enough, many of these symptoms are most apparent in the biomedical sciences,[10] even though it is precisely these fields to which the largest volume of federal funding has been devoted.

Some Possible Explanations

How might we understand the causes of these symptoms of persistent malaise in what is otherwise a heavily funded and highly productive research system? Are they likely to persist, worsen, or diminish in the foreseeable future?

The overarching reality is that federal funding for research is concentrated in the "discretionary" portion of the federal budget (as are other key sectors such as defense, education, transportation, etc.). It is this discretionary part of the budget that now is being squeezed by the automatic expansion of nondiscretionary spending in a context of slow economic growth and slow-growing federal revenues. Nondiscretionary spending includes so-called entitlements such as Social Security and Medicare, along with the interest payment obligations of large and rapidly expanding government debt.

Of course it is possible to imagine political and economic developments that would reduce the current pressures on the federal government's discretionary budget. The difficult bipartisan political decisions that would be needed to restrain continuing expansion of nondiscretionary expenditures might be adopted. Overall federal revenues might be increased through some combination of tax increases and faster economic growth. The federal government might find it possible to continue to increase its debt so as to enable expansion to continue in both the nondiscretionary and discretionary budgets.

While some or all these possibilities are possible in theory, it is only fair to say that all appear rather unlikely in view of the prevailing economic and political circumstances, not to mention the constraints imposed by the already large burden of federal debt accumulated over the past decade alone. Still, if we were to assume that a combination of these favorable scenarios will emerge in the short- to medium-term future and be of sufficient magnitude, actions to address current difficulties could be at least deferred even if the underlying structural challenges remain.

The discussion that follows does not embrace that comforting assumption. Instead it is based on assumed outcomes that appear to be far more likely: that growth in federal research funds will remain constrained for the foreseeable future, and that the realities of current severe pressure upon federal discretionary expenditures will not disappear. Readers who believe otherwise can ignore much of what follows.

The current symptoms of malaise previously described do not result from declines or sustained low growth in federal funds for basic research over the past decades. To the contrary: federal research funding over these periods has expanded very substantially—to be sure both unevenly across fields, and fitfully via a series of booms and busts—but nonetheless expanded.

Figure 8.1 shows the long-term trajectory of overall federal R&D outlays from 1949 to 2014, expressed in constant dollars to adjust for overall inflation. The data are broken down between defense and nondefense, lest there

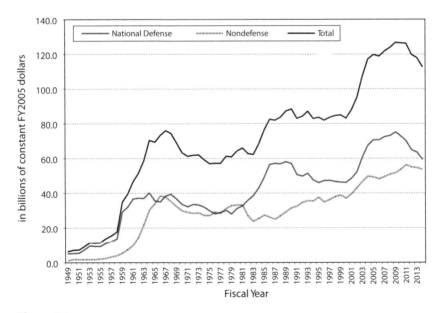

Figure 8.1.
Federal research and development outlays, total, defense, and nondefense, in constant FY2005 dollars, FY1949–FY2014.
Source: Figure prepared by author from data obtained from White House, Office of Management and Budget, President's FY2014 Budget, Historical Tables, table 9.7. Available online at http://www.whitehouse.gov/omb/budget/Historicals.

be distortions in total federal support for R&D due to expanded or reduced defense budgets driven by international developments. As noted in chapter 7, federal outlays represents less than one-fifth of total U.S. R&D, but account for the majority of basic research.

These graphs illustrate the sharpness of the federal R&D funding boom over the decade beginning in the late 1950s (i.e., after the political shock of Sputnik), followed by a significant decline and then by stagnation in federal funding (adjusted for inflation) from the late 1960s to the early 1980s. Then, from the early 1980s to around 1990 there was a second funding boom, followed by another period of slow growth or decline in R&D outlays for nearly a decade, adjusted for inflation. During the years around the turn of the twenty-first century, there was a third federal funding boom, driven mainly by the rapid doubling of the NIH budget and the expansion of defense budgets related to the September 11 attacks and the wars in Iraq and Afghanistan. There ensued a few years of decelerating R&D increases followed by another brief surge of research funding including part of the economic stimulus programs, and then another substantial decline in constant dollar terms.

It would difficult to conclude from figure 8.1 that the malaise symptoms can be attributed to declining or even slow-growing federal support for nondefense R&D. The funding shifts underlying the boom and bust cycles discussed in earlier chapters are readily visible, but the overall trend has been one of substantial increase that is well in excess of inflation.

If persistent underfunding is not the cause of current symptoms of malaise, what might it be? The most likely candidate is a combination of the erratic and unpredictable booms and busts of federal support for R&D, coupled with a set of perverse and unintended incentives and positive feedbacks that have evolved in the U.S. research system. Together these conspire to subject the system to damaging instabilities and increasing signs of malaise, in spite of substantial funding increases over the past decades.[11]

Careful analyses of these unintended but perverse incentives and positive feedback may provide a road map of incremental adjustments that could guide the system toward a more stable and sustainable equilibrium, while avoiding damage to its very high quality and its important public-good contributions to human welfare. Indeed, any significant changes should be sensitive to the system equivalent of the Hippocratic Oath: treatments to mitigate negative symptoms should, above all, "do no harm."

Perverse Incentives and Positive Feedback: Unintended and Destabilizing

The characteristics of the current system that are damaging and destabilizing were not intended by anyone.[12] Instead they have evolved over time to become part of the current system, without intent or plan.

Federal appropriations are annual affairs, but creation of research infra-structure and human capacities for leading-edge research requires lengthy periods of sustained commitment. The long history of boom-and-bust, of feast and famine, in government financing for science and engineering has been particularly damaging to activities that require long lead times, quint-essential characteristics of basic research and higher education. The damage falls most heavily upon four categories that are central to the U.S. research system: recent PhDs and postdocs, newly hired faculty members, established research faculty and laboratories, and research institutions:

- *Recent PhDs and postdocs*: As we have seen, under the current system a sudden acceleration in research funding can be expected to expand the numbers of PhD students and postdocs, primarily in their roles as research workers financed by federal grants. After a lag of the multiple years that are required for PhD and/or post-doctoral completion, these expanded numbers of recently qualified junior researchers emerge into the labor market in pursuit of ca-reers in research. If by then the increase in funding has decelerated (i.e., the boom has either waned or become a bust), the outcome is poor career prospects for young researchers—some of whom are truly outstanding by any measure—who have invested many years of their lives and large amounts of government, institutional, and personal resources to qualify for careers that are no longer readily available. Some of those affected resort to extended temporary em-ployment as long-term postdocs. There is evidence that many in such positions perceive personal failure and discouragement, and some express anger at having been misled by "the system." Many eventually depart from the research careers for which they spent so many years preparing.

- *Newly hired faculty*: Research funding booms often lead to ad-ditional faculty hiring by universities and research institutes. Then as the boom wanes, all researchers, but especially more junior in-vestigators, find it increasingly difficult to obtain research funding. The unintended and damaging effect is to disrupt promising re-search careers. To the extent their remuneration depends upon ex-ternal grant funding, even truly outstanding junior researchers may face serious personal financial challenges not of their own making.

- *Established research faculty*: As funding booms wane, success rates decline for proposals submitted even by outstandingly accom-plished senior researchers. Understandably this generates strong incentives for them to devote increasing fractions of their time and creative energy to preparing additional funding proposals. The expansion in proposal numbers in turn results in more time and effort spent by researchers and federal funding agencies in implementing peer review. These effects may be magnified if funders adopt the expedient of shortening the length of funded grants to spread constrained funds more widely. Moreover, even highly productive researchers with established records of research success may experience lengthy gaps in renewal grant support, thereby disrupting already successful research careers, imposing stresses upon their institutions, and sometimes even leading to closure of productive research labs.

• *Research-intensive institutions*: During episodic booms in federal research funding, administrators of some research-intensive institutions may respond to incentives to "leverage up" by expanding their faculties or research facilities, as discussed in greater detail in the following sections.

Perverse Incentives Favoring Increased Leverage

The current system, again unintentionally, provides incentives that encourage research-intensive institutions to increase their financial leverage and hence their vulnerability. These incentives operate both on faculty salaries and on the financing of research facilities. So long as federal research funding increases robustly this strategy appears quite appealing, but when funding booms wane, such institutions may face serious problems in covering expanded but unfunded faculty salaries and in servicing the debt incurred for expansion of their research facilities.

These problems have been magnified by recent parallel declines in other sources of institutional revenue (from endowments, from state contributions to public research universities). One effect—a doubly perverse one—has been to increase even further institutional incentives designed to maximize success in obtaining federal research funding. Another has been to discourage institutions from making longer-term commitments to tenured or tenure-track faculty, thereby generating another positive feedback loop in which they rely increasingly on less expensive and inherently temporary research workers such as graduate students and postdocs.

Regarding salaries, the largest federal research funder, the National Institutes of Health, places no limits on the percentage of faculty salaries that can be covered by its grant funds.[13] Payment of large fractions of faculty salaries appears to be most common in medical schools and in freestanding nonprofit research institutes, and less common in colleges of arts and sciences where tuition and teaching constitute a larger part of the revenue stream and faculty effort in science departments.[14] In contrast, the National Science Foundation has long limited salary payments for senior research project personnel to no more than two months of their regular annual salaries.[15]

Understandably, those senior faculty whose own salaries and benefits depend heavily upon external grant funding experience strong incentives to commit time and effort in pursuit of new and renewed external research funding, not only in support of their research interests but also of their own personal compensation. They also face greater risks of funding gaps that threaten research programs.

The same incentives may lead administrators of their institutions to encourage faculty to maximize the proportion of their salaries and benefits paid by external but ultimately "soft" funding sources such as NIH research grants. These sources of funding include an added incentive for administrators, in that salary reimbursement from grants is included in the base for negotiating indirect cost rates. The odd and unintended outcome has been described in the following declarative way by Bruce Alberts, former president of the National Academies and currently editor of *Science* magazine:

> NIH actually rewards institutions for paying faculty salaries with unguaranteed "soft money" from research grants by providing increased overhead payments. Amazingly, any institution that draws on its own finances to pay its professors is doubly disadvantaged: It must not only use its own funds but also loses the overhead on the salaries that it would otherwise accrue.[16]

In effect these incentives have ineluctably led medical schools and other institutions to become financially more leveraged, by encouraging them to become increasingly dependent upon external grant funding to cover their faculty payroll. In doing so, faculty positions (including even those protected by tenure) and labs become increasingly unstable if proposal success rates drop because of the waning of funding booms.

In response, university administrators emphasize that any NIH limitations on the percentage of investigator salaries payable under NIH grants would make participation in NIH-supported research financially more difficult for many institutions. They note also that the issues are exacerbated by recession-related funding pressures on clinical revenues, endowment income, charitable donations, and state support. Their concerns about all of these trends are well-founded.

Regarding research facilities, similar incentives are in place that favor increased financial leverage. Current federal regulations (codified in OMB Circular A-21[17]) encourage research universities to borrow funds for building, renovation, or re-financing of research facilities, by allowing universities to include the costs of debt service in their indirect cost calculations for grants, but prohibiting them from doing the same for the (imputed) value of similar laboratory facilities that are not debt financed.[18]

Presumably unintentionally, these incentives built into the current funding system encourage institutions to fund faculty salaries and research facilities on federal grants, thereby rendering them more dependent upon continuing increases in federal R&D funding and more vulnerable if such increases do not continue. As noted in the next section, this structure also implies that acceleration in grant funding will likely lead to trends that in turn have their own feedback effects on the future supply of U.S. graduate students and postdocs.

Positive Feedback with Limited Information

In addition to these perverse incentives, destabilizing positive feedback is a characteristic of the current system as it is structured. As discussed earlier, a boom in research funding will predictably generate recruitment of more PhD students and postdocs, increasingly imported from abroad (see chapter 6), to serve as research workers on the additional or expanded research projects. Yet if and when the research funding boom wanes, the expanded number of PhD graduates and postdocs who will complete their training are likely to face chilly career prospects.

As noted earlier, in a normal market such a system would gradually self-correct, albeit with real pain for those involved in the correction. Prospective new entrants to the PhD or postdoc would have access to information about the poor career prospects, and over time fewer entrants both from the United States and from abroad would pursue these careers. However, as we shall see, this particular system does not have effective mechanisms for such self-correction.

Though it may be surprising to some, most graduate departments in science and engineering provide prospective new PhD students and postdocs with little or no credible information about the career outcomes their past PhDs and postdocs have experienced. They may report on the postdoctoral or other initial positions to which their PhD graduates have gone, but say little about what former PhD students and postdocs are doing five or ten years after they departed. This situation is notably different from the practices of law or business schools at the same university, which typically do report publicly on the career experiences of their graduates.[19]

The result is that many prospective new graduate students and postdocs are being offered substantial financial inducements to enter a lengthy PhD/postdoc sequence while lacking realistic understanding of their prospects in the research labor market beyond the PhD/postdoc, and particularly their prospects for obtaining a tenure-track research faculty position. Indeed, the ready availability of federal financial inducements may send (again unintentionally) a signal to prospective PhD students and postdocs that the federal government is investing in their advanced education because it anticipates there will be strong labor market demand for the research knowledge and skills they will develop.

The non-alignment of demand with supply of completing PhDs and postdocs also likely contributes to one of the malaise symptoms discussed earlier—the lengthening of the PhD and postdoc training. If career positions are scarce but funding for temporary positions such as graduate research assistants and postdocs is ample, aspiring PhD students and postdocs understandably may choose to extend their training for another year or more. They do so in hopes that career conditions will improve in the interim, and also that they will be able to enhance their career prospects by successfully publishing their research. Meanwhile their mentors may experience congruent incentives for lengthening the terms of their PhD students and postdocs, because research assistants who stay longer have more experience in a particular lab and research area and hence may be more productive in conducting the funded research. Requiring research assistants to focus narrowly on their research topic may also maximize the lab's research productivity, and yet at the same time constrain the breadth of their training—thereby (unintentionally) limiting their career prospects even further.

The results of all these positive feedbacks and perverse incentives are not only destabilizing to the research and education system, but also produce frustration and distrust among those who feel they have been misled. This may provide impetus toward the unionization of graduate students and postdocs, a trend opposed by most academic researchers as antithetical to high-quality education.[20] It may also cascade down the higher education system to the level of talented undergraduates interested in science and engineering, as the PhD students and postdocs who serve as teaching assistants for undergraduate courses at many research-intensive universities communicate their own discouragement.

Finally, there may also be unintended effects upon the type of person attracted to research careers, if talented individuals who have the greatest range of opportunities or aspire to a certain standard of living opt out of research in favor of other careers.

Looking Forward: Candidates for System Adjustments

The United States can justifiably be proud of the outstanding research enterprise it has created since World War II. It has produced major advances in scientific understanding and economic productivity, and has also contributed to splendid scientific breakthroughs and technological innovations that have improved the well-being of hundreds of millions around the world. Simply put, this is a system that amply justifies continuation of substantial government support.

Past "wolf at the door" claims about declines and shortages in U.S. science and engineering have proven to have been overstated. Yet as noted in chapter 7, this does not mean that another wolf might not be at the door in the future. To the contrary, without participating in the decades-long practice of overstatement to gain political traction, we can see that the U.S. system that supports both spectacularly successful research and the advanced education of the future science and engineering workforce has been showing proliferating signs of increasing instability and malaise. These will need to be addressed if the current high levels of quality and productivity of the U.S. science and engineering enterprise are to be sustained into the future.

There is no need for revolutionary change, no necessity for root-and-branch overhaul. Instead, what is needed is judicious application of incremental adjustments to the current structure of incentives and disincentives, many of which were never intended to have the effects they appear to be causing. The ambitious goal would be to moderate current incentives that are problematic, and to maximize those that serve to make the system healthier, more stable, and more productive. Applied intelligently and carefully, such adjustments also could gradually make research careers more attractive to talented and capable students. The closing discussion addresses some of the candidates for such adjustments.

Identify Sensible Ways to Moderate Future Booms and Busts in Research Funding

As we have seen, rapid acceleration and deceleration of funding for scientific research can unintentionally inflict real damage upon the U.S. research system. The series of boom/bust cycles since World War II have in many cases represented responses to arguments that the United States is falling behind its competitors or facing "shortages" of scientists and engineers, thereby posing serious threats to national security or economic competitiveness. These alarms have been echoed by mainstream media and embraced by many leaders in business, politics, and education.

Objective researchers have not been able to find much evidence of such generalized and widespread shortages in the data on the U.S. labor market for scientists and engineers. However, concerted lobbying based on these claims has been impressively successful in political terms. The resulting policies have led to large but erratic increases in both federal research funding and temporary visas. These in turn have inflicted their own destabilizing damage to the system. This is a fundamental problem of U.S. politics, controlled as it has increasingly become by heavily funded and energetic lobbying and advocacy.

Of course it is always impossible to be sure about what historians call "the counterfactual"—what would have happened if the alarms had not been repetitively sounded about workforce shortages leading the United States to fall behind its competitors. Yet it seems quite likely that the U.S. research system would be far healthier now if the same funding increases injected over the past two decades in response to such alarms had been provided in a more gradual but predictable way, and if expanded visa provisions had been better conceived and implemented. With this in mind, it would be both wise and refreshing if future advocates would emphasize the importance of more steady and predictable funding increases, rather than rapid acceleration that typically has been followed by rapid deceleration. Similarly it would be constructive if future visa policies could be based upon credible evidence and analysis by objective analysts, rather than on questionable claims by parties-at-interest.

Prior experience is not encouraging about the likelihood of such changes. Over the past two decades well-intentioned advocates for expanded federal support for research have repeatedly pressed for the arbitrary goal of a "doubling" of federal funding over a short period of years. The doubling of the NIH budget from 1998 to 2003 provides a valuable natural experiment. As we have seen, shortly after these rapid funding increases ended, the biomedical sciences were battered by a new "funding crisis" of serious magnitude that continues to reverberate.

The initially positive, but ultimately destabilizing, effects should provide sufficient warning about future efforts. If the politics are such that the most compelling lobbying position is to seek another budget "doubling," it would be far healthier for all concerned if such a 100 percent budget increase were to take place over a more extended number of years, for example, ten to fifteen years rather than the five years of the 1998–2003 NIH doubling. And no one should imagine that doubling of research budgets can resolve the structural problems that underlie the instabilities of the system over the past decades.

Level the Playing Field of Incentives, and Reduce Destabilizing Feedback

The health of the current U.S. science and engineering system also could be greatly enhanced by careful reassessment of the array of incentives that undergird it. Many of the effects of these incentives were neither intended nor anticipated at the time they were implemented. Moreover, most were put in place decades ago, when circumstances were quite different. If such reassessments were to conclude that some or all of the current incentives continue to have merit, they nonetheless could constructively be gradually adjusted to enhance the system's ability for nimble self-correction as external circumstances change. Adjustments that may be worthy of attention include the following.[21]

1. Reduce or eliminate the incentives incorporated into OMB Circular A21 that encourage research institutions receiving federal research grants to finance their research facilities via debt rather than equity ownership.

2. Incrementally apply limits to the percentage of PI salaries payable under NIH research grants. The absence of such limits has also incentivized institutions to "leverage up" by using debt to expand their research facilities and faculty, in anticipation that this debt can be financed by indirect payments from expanded NIH grant funding. (NSF rules do have such limits and hence do not provide the same incentives.) One respected scientist-statesman has called for NIH to "make it unambiguously clear that expansion through laboratory building construction requires a substantial, nonreimbursable, long-term commitment of resources, including 'hard-money' faculty support by any institution that wants to increase its facilities and research staff."[22] A similar recommendation was included in the 2012 report of the Biomedical Workforce Working Group of NIH's Advisory Committee to the Director.[23]

3. Improve career information available to prospective new PhD students and postdocs by requiring departments and institutions receiving federal research funding to provide prospective applicants with web access to accurate data on the early- and mid-career experiences of their past PhDs and postdocs. This recommendation was also embraced by the NIH Biomedical Workforce Working Group.[24] Additional web-based information on outcomes of PhD education and postdocs could be provided by federal funding agencies themselves, including links to other sources such as scientific society websites. Such web resources might even include a career clearinghouse for researchers.

4. Limit the years spent in PhD study and postdoctoral positions, thereby enabling the system to respond more nimbly to changing demand. The NIH Biomedical Workforce Working Group recommended that a cap of five years be adopted for support by NIH funds of any type (training grants, fellowships, research project grants).[25] However, this would have to be done carefully lest shortening the

time-to-PhD have the perverse effect of increasing the number of PhDs awarded in ways that would further diminish the attractiveness of pursuing doctoral education.

5.   Clarify the goals of using federal research funds to finance unlimited and increasing numbers of international PhD students and postdocs, for whom opportunity costs are lower and stipends and prospects of permanent visas are more attractive than for their U.S. peers. Is the main goal:

• to increase the size of the U.S. science and engineering workforce?

• to lower research costs by staffing federally supported research labs with poorly paid research assistants?

• to create international research connections, or to enhance the research capacity of their countries, if and when they return home?

6.   Modulate the predictable expansion of PhD students and postdocs resulting from increased research funding (even if faculty and administrators recognize that the career prospects for these additional numbers are not favorable), by reducing the dependence of university research laboratories upon large and increasing numbers of temporary PhD student and postdoc research assistants as their workforce. This could be done by encouraging institutions to create more permanent PhD-level laboratory staff ("staff scientists"), by encouraging NIH and NSF peer reviewers to be receptive to grant applications that include them, and by ensuring that PhD students and postdocs receive stipends and benefits commensurate with their levels of education rather than continuing current practices that (unintentionally?) make them much less expensive as research workers.[26]

The primary issues of this kind relate to the doctoral level, upon which most federal funding for graduate education has come to focus. Yet most of the science and engineering workforce has lesser levels of education, and non-academic employers claim they find it difficult to recruit scientists and engineers who have both the technical skills beyond the baccalaureate level and the nontechnical skills and experience required for success in high-tech business careers.

Federal funding agencies interested in the adequacy of the U.S. science and engineering workforce could diversify their support by encouraging graduate degrees other than the PhD that are more responsive to the changing needs of non-academic employers. One promising example is the Professional

Science Master's (PSM) degree that has been rapidly gaining popularity around the country.[27] These two-year graduate degrees have been configured by science faculty members to educate science professionals for industry and government. They are advised by *non*-academic employers seeking to recruit science professionals who combine graduate-level knowledge of scientific and technical disciplines with basic knowledge of nontechnical subjects that are necessary for success in business—management, finance, teamwork, communication, legal and regulatory affairs where appropriate, and so on.

PSM degrees are quite new by the standards of graduate education, with the first few such degrees emerging only fifteen years ago. Since then they have been expanding rapidly in response to strong interest from both employers and students, and have been enthusiastically endorsed in reports and assessment from a wide range of leading organizations in science (e.g., two committees of the National Research Council[28]); business (e.g., the Business Higher Education Forum and the Council on Competitiveness[29]); education (e.g., the Council of Graduate Schools[30]); and government (e.g., the National Governors Association and the National Conference of State Legislatures[31]). Degrees of this type received limited support in the past from the National Science Foundation, and have been endorsed by NIH's Biomedical Workforce Working Group, but currently receive little or no federal support.[32] By 2013, the number of such degree programs exceeded 300 at more than 135 U.S. universities,[33] thereby demonstrating substantial responsiveness on the part of U.S. higher education to national workforce needs.

Create a Mechanism to Provide Objective Assessments of Career Trends and Claims of "Shortages"

Given the importance of the health of the U.S. science and engineering enterprise and its talented workforce, it is striking that there continues to be no credible entity charged with regular, systematic, authoritative, and objective assessment of trends and prospects for education and careers in these occupations. Those making public policy decisions relevant to the science and engineering workforce are continuing to fly blind, rather as if federal budgetary decisions were being made with no inputs from the Office of Management and Budget and the Congressional Budget Office.

Much of the running to date has been made by advocacy groups, often using data of highly questionable validity. As discussed in chapter 3, one of the most effective rhetorical weapons in advocacy campaigns have been claims of current (or sometimes "looming") shortages of scientists and engineers. Some of those who express concerns about shortages do sincerely believe they exist. Others may have doubts but are following the advice of lobbyists who tell them that shortage claims offer the most effective arguments in support of the goals they seek to achieve.

A number of models have been put forward to create such a mechanism, most as a result of debates about visa policies for skilled personnel. None of these proposed models has been adopted, but this is not an issue that can or should go away. Ideally such an effort would engage labor market experts selected on the basis of their technical knowledge rather than their political connections, and would operate independently of funding agencies that may have conflicts of interest. A number of candidate mechanisms warrant attention here, with brief comments offered as to the pros and cons of each.

- The National Science Foundation (NSF) could be charged with this task. Its National Center for Science and Engineering Statistics (NCSES)[34] has established a reputation for high-quality surveys and other data collection efforts on the science and engineering workforce. In an ideal world this NSF Center would be able to conduct independent and sophisticated assessments of shortage claims. *However*, the NSF has its own institutional interests in expanding its budget and so is not seen as wholly independent. It has not been forgotten that in the late 1980s the NSF leadership— under different management to be sure—promoted dubious claims of "looming shortfalls" of scientists and engineers. (For more details on this episode, see section entitled "'Shortfall' Studies at the NSF" in chapter 2.) The 2012 report by NIH's Biomedical Workforce Working Group recommended that NIH also create "a permanent unit in the Office of the Director … to coordinate data collection activities and provide ongoing analysis of the workforce and evaluation of NIH policies so that they better align with the workforce needs,"[35] a recommendation that has been welcomed by the NIH leadership.

- The Bureau of Labor Statistics (BLS) in the Department of Labor is widely respected for its fierce independence from political pressure, an independence that is essential given that the BLS reports on politically sensitive questions such as the rates of unemployment and inflation. In addition to reporting on labor market conditions, the BLS already produces a regular series of occupational projections described elsewhere in this volume. *However*, the data and analyses produced by the BLS address the entire U.S. labor force, of which the science and engineering components represent only very small fractions. The BLS has no particular interest or expertise in dealing with the special features of these relatively small labor markets, although it could develop such expertise if charged with this responsibility.

• A third possibility would be a specialized and independent analytic group focused precisely upon the science and engineering labor market, located at a high-quality university or research institute. One candidate would be the Science and Engineering Workforce Project (SEWP) at the National Bureau of Economic Research (NBER).[36] *However*, SEWP is currently a rather small research operation with no permanent staff, more like a loose network of labor economists, and hence would require a substantially different structure and stable funding sources in order to take on such a new set of roles.

• A fourth possibility would be to place these science and engineering workforce issues on the agenda of one think tank's proposal for a "Standing Commission on Labor Markets, Economic Competitiveness, and Immigration."[37] This idea has the virtue of independence from any federal agency that might have an interest in the outcomes, and as proposed would force action by an otherwise stalemated Congress. *However*, members of such a Standing Commission, if appointed politically as proposed, would most likely not be truly independent of the views held by those who appointed them, and the Standing Commission as a whole would represent a high-value target for capture by all of the many interest groups that surround this subject.[38]

• Finally, a somewhat different approach could be based on an experiment under way in the United Kingdom in the form of its independent Migration Advisory Committee (MAC). This model differs substantially from the "Standing Commission" outlined earlier in that its members would be selected from applicants using external peer review, and its analyses and recommendations would be advisory only. *However*, this UK committee addresses labor market questions only as they relate to the immigration of skilled persons, although a similar U.S. committee could be charged with assessing the entire science and engineering labor market and workforce. In addition, the UK system functions in the context of a parliamentary system in which the "government" is controlled by the parliamentary leadership and the civil service is granted more discretion than in the U.S. system. (The UK committee is described in greater detail in the sidebar.)

Some Watchwords for the Future: A Guide for the Perplexed

There are a number of principles and cautions that warrant attention in future discussions of this topic.

First: "Do no harm." The spirit of the Hippocratic Oath should apply to proposed changes in a research system that has been highly productive. Of course doing nothing can also do harm, as surgeons and central bankers would attest. "Do no harm" should not be a recipe for immobility, since systems that do not adjust to changing circumstances tend to founder.

Second: Avoid any sustained periods of imbalance between supply and demand in the U.S. science and engineering workforce. Short-term imbalances are manageable via quite feasible adjustments in remuneration, the organization of work, or internal and international migration. Long-term imbalances are far more disruptive, however, and require careful analysis to identify smart interventions that do not create their own damaging dynamics.

Equal attention should be paid to minimizing both under- and over-supply. It is obvious that substantial under-supply relative to demand over a period of many years can constrain feasible science/engineering advances and their humanitarian and economic payoffs. Meanwhile some are fond of arguing that substantial over-supply of scientists and engineers relative to demand actually is benign ("one can never have too many scientists and engineers" who will "find their way"), but this is a shortsighted perspective. The realities are that education for careers in science and engineering requires large investments of time and resources, and that a situation of sustained over-supply likely deters new entrants by undermining the attractiveness of these careers for talented students, thereby sowing the seeds of subsequent under-supply.

Third: Avoid rapid acceleration and rapid deceleration in funding—in Vannevar Bush's words, seek "stability of funds over a period of years." The U.S. research system is large and complex, and requires long-term commitments to continue its high level of quality and productivity. Experience over the past half-century has amply demonstrated its vulnerability to boom-and-bust funding patterns. While it has proven attractive to many supporters of science to push for rapid funding increases—the preferred target seems to be an arbitrary "doubling"—the evidence indicates that advocates should be careful what they wish for.

Fourth: Anticipate the unintended. The past half-century in this area is littered with the casualties of unwanted and unexpected consequences of research funding and policy initiatives. We have seen how the 1998–2003 doubling of the NIH budget has produced its own set of unanticipated "crises." Nor did anyone anticipate that the combination of increasing government research grant funding, the rules on their use by grantees, and the imbalances of the U.S. visa system would together result in *research* dollars coming to dominate support for graduate study in science and engineering, and later would lead to more than half of biomedical postdocs represented by PhDs from abroad.

Fifth: Beware long-range "forecasts." Many of the claims of "shortages" of scientists and engineers have been "looming," that is, based on forecasts (or,

more accurately, projections that are put forth as forecasts). It is embarrassing to note that the history here is the history of the graveyard—when compared with subsequent developments, almost all such efforts have proven
to be failures, as illustrated in the summaries in chapter 2 of such efforts by
Brode, House, Bowen and Sosa, ITAA, and others. A 2000 report by a high-
level National Research Council workshop concluded "that accurate forecasts have not been produced"[39] and recommended action to improve both
the data and methods that would be needed. Regrettably none of this has
been done, and until it is, real caution is in order for the use of projections
of demand and supply related to the science and engineering workforce.
Although properly formulated projections do provide useful forward looks
over the short term (e.g., 2–3 years), they also demonstrate that the range of
uncertainty becomes very wide in the medium-and-longer term.

Reality Check

Reality checks are appropriate for all of the previous suggestions, as it may
be that some or all of the approaches described are not politically feasible.
The current structure may simply be too deeply entrenched and powerfully
resistant to change, and those who benefit most from it may be content to
continue the past cycles of boom and bust, feast and famine, and the uncertainties and perverse outcomes these produce.

   The most concise way to test the reality value of some of the suggestions
in this chapter might be to challenge them with a series of hard and skeptical questions:

   1.   Even if it would be highly desirable for science and engineering to have steady and predictable growth in federal support
   rather than the booms and busts of the past, wouldn't the annual
   federal budget and appropriations cycle make this desirable goal
   impossible?

   2.   Shouldn't the alarms about "shortages" and crises, even if they
   may have proven exaggerated, be treated simply as necessary evils
   required to capture the attention of those responsible for relevant
   policy?

   3.   Would the U.S. Congress welcome a nonpolitical and techni-
   cally sophisticated advisory entity to assess claims of STEM workforce shortages?

   4.   Would the numerous advocacy-oriented groups that long
   have stalemated any action on "skilled" visa changes be willing to
   compromise?

5. Would the NIH be willing to incur the wrath of its support network of research universities and faculty by limiting the percentage of faculty salaries that can be paid from NIH grants, the number of years NIH funding can be used to support graduate students and postdocs, or the number or growth of international PhD students and postdocs supported under NIH research grants?

6. Would research institutions be willing to pay for more expensive and longer-term staff scientists instead of continuing the current dependence upon shorter-term and less expensive PhD students and postdocs?

7. Would senior and well-funded investigators support measures that would improve the funding prospects of junior investigators when funding is tight for everyone?

Some of these challenges may lead one to conclude that any effective adjustments to the structure of the current system are not likely. If so, we should expect a continuation of the current symptoms of malaise for a period of years as the fortunes and federal funding for science and engineering fluctuate. The most likely outcome in such a scenario will be a succession of further booms and busts, perhaps again stimulated by alarms that are designed to capture the attention of political leaders.

In this scenario, a succession of booms and busts will continue to flow back and forth over the system like inexorable tides, but tides that inflict serious damage by being impossible to anticipate.

The United Kingdom's Migration Advisory Committee (MAC)

The UK Migration Advisory Committee (MAC) is a technical group established in 2007 by the then-Labour government. It emerged primarily due to earlier policy failures that had compromised the credibility of the British government's own analyses of immigration patterns and potentials.

The volume of immigration to the UK increased rapidly during the 1990s, but was accompanied by increasing public and political concerns about failure in the integration of immigrants. Such concerns came to an explosive head with the appearance of "homegrown" Islamic militants, some of whom committed terrorist bombing attacks in the London Underground and elsewhere. Meanwhile, the UK government's policies regarding refugees and asylum were severely criticized as "lax."[40]

The final straw, though, was the UK government's failure to anticipate the very large inflows of legal migrants who arrived in the UK from the eight "new accession" countries, mostly in Eastern Europe, that were admitted to the European Union in 2004. Because the relative living standards in many of these new EU members were substantially lower than those of the existing seventeen EU members, fourteen of those seventeen countries imposed multiyear periods in which there would be numerical limitations upon labor migration from the new EU member countries on grounds that time must be allowed for greater convergence of economic conditions. In contrast, the governments of three of the seventeen EU member countries (the UK, Ireland, and Sweden) decided to allow unlimited labor migration from the new EU member countries.

In the case of the UK, this government decision was based upon what in retrospect were its badly flawed assumptions as to the likely volume of such migratory movements. In public debate before it implemented its "no-delay" policy, the UK government stated that only a small number (5,000–13,000) of such migrants would enter the UK by 2008. In the end some 765,000 migrants from the accession states entered the UK, at least on the basis of official registration numbers (i.e., nearly 60 times higher than the government's projection). The actual total of such migrants including those who did not register may have been larger, at about 1.5 million, which would represent more than one hundred times greater than the official estimate.[41]

The wildly low estimates on which the UK government based its position generated animated and widespread public controversy and criticism, and inflicted serious damage upon the Labour government's credibility on immigration policy. In response to such criticisms—and perhaps also due to simple embarrassment—the Labour government established the new Migration Advisory Committee as an independent committee of five prominent independent labor market experts. Unlike most U.S. government advisory committees, its members were selected by a peer review (rather than a political) process from a pool of applicants who responded to a public announcement. The members serve on a part-time basis with support from a small technical staff.

The MAC is independent, but it is not an independent "commission" of the type being advocated by some in the United States. Its responsibilities are to advise the British government rather than to put forward policy changes that require acceptance or rejection by the government. It responds only to official government requests for advice, and does not initiate its own studies.

The MAC appears to be entirely nonpartisan, and to be seen as such. Indeed, when the Labour government was defeated by a coalition of the Conservative and Liberal Democrat parties in the elections of 2010, the new coalition government continued the MAC and made no changes in its membership.

The MAC summarizes its approach with three "S" words—Skilled, Shortage, Sensible. It responds to requests from the government for an objective assessment of claims (usually from employers) that they face workforce shortages in specified skilled occupational categories, for which they are seeking the admission of more workers from abroad. The MAC first determines if the occupation in question actually is a "skilled" one; it does not deal with unskilled occupations. If the occupation does meet the criteria as skilled, the MAC proceeds to develop a wide range of quantitative data and other evidence to assess whether the claimed "shortages" are real. Finally, if it determines that real shortages do exist, it analyzes whether immigration is a "sensible" response when compared with other approaches such as expanded training programs or appropriate increases in remuneration.

The MAC has developed a rather sophisticated methodology that combines or "dovetails" high-level quantitative data (usually from official government sources, which it terms "top-down" data) with qualitative and more detailed insights solicited from stakeholders in specific industries, occupations, and regions, including employers, industry organizations, and unions or professional societies where appropriate ("bottoms-up"). The MAC reports that the top-down data are essential to its recommendations, but also that the bottoms-up inputs it solicits "form an integral part of our

analysis. This evidence enables us to examine individual occupations and job titles with a degree of focus that is not possible
through top-down analysis alone."[42]

The MAC's findings and recommendations to the government are
published and are widely reported in the press, but they do not
determine policy. Instead, the government reviews the MAC's
recommendations and decides whether it wishes to implement
some or all of them, and announces its decisions. The process is
open and public, and is generally seen to have raised the quality
of debate about what previously were decisions taken behind
closed doors on purely political grounds.

Among the alternatives, a U.S. adaptation of the model pioneered
by the MAC may have most promise of providing the government and the public with objective analyses of claims that there
are current or "looming" shortages of scientists and engineers.
Whether development of such an independent expert entity is
feasible in the United States given its political structure and
interest groups is another story.

APPENDIX A
Controversy about the Meaning of Sputnik

There is an interesting controversy about the background of the Soviet success in beating the United States into orbit with its Sputnik series of satellites. A 2004 article by T. A. Heppenheimer[1] makes the challenging argument that the U.S. government deliberately allowed the Soviets to launch the first orbital satellite. Heppenheimer's argument in brief is as follows:

For nearly a decade, the U.S. military had been secretly exploring the feasibility of a then-fanciful idea, being developed by RAND and other think-tanks, of a satellite surveillance system that would be able to prevent a surprise nuclear attack from the Soviet Union. In March 1954, RAND had produced a classified two-volume report of the studies on this system that it had undertaken under this "Project Feedback." However, RAND analysts pointed out that in the absence of any man-made satellites there was no legal precedent or principle by which it could be established that such surveillance of Soviet territory from orbit would not be seen as an infringement upon sovereign Soviet airspace—as high-flying surveillance aircraft such as the U-2 clearly were. Heppenheimer states that proponents of this satellite surveillance system therefore urged that the United States allow the USSR to be the first to orbit a satellite as part of the International Geophysical Year (IGY), thereby itself establishing the precedent that overflights of other countries' territories from space did not represent infringement of sovereign airspace.

It is in large part for this reason, Heppenheimer argues, that the U.S. government ordered the Army's rocket program led by Wernher von Braun to halt any effort to develop a launch vehicle for the IGY capable of achieving Earth orbit. In addition, there was concern about Soviet perceptions of a scientific satellite launched for the International Geophysical Year by a military rocket program led by a former Nazi scientist. Instead, the Naval Research Laboratory, well known for its geophysical research, was chosen to develop the Vanguard scientific satellite program, based upon existing rockets that previously had been used for such scientific research rather than military purposes.

At that time, the army had already developed its Jupiter-C rocket based on the existing Redstone rocket. The Jupiter-C was far more powerful than the navy's Vanguard program, and indeed the calculations showed that it needed only the development of a fourth stage to be able to put a small satellite into orbit. Wernher von Braun was anxious to do so, but the Heppenheimer article quotes a specific phone call to him from Major General John B. Medaris, Commander of the Army Ballistic Missile Agency (ABMA), and von Braun's direct supervisor. In this September 20, 1956 call, more than one year before the Sputnik launch, Medaris is quoted as follows: "Wernher, I must put you under direct orders personally to inspect that fourth-stage to make sure it is not live."[2]

Von Braun followed Medaris's orders, and according to Heppenheimer this strategy to slow the U.S. effort so that the USSR would beat the United States into orbit worked as planned. By launching Sputnik 1 into Earth orbit, the Soviets established the international legal principle that orbiting satellites did not infringe on sovereign airspace, and this then enabled the United States to develop and deploy its ambitious space-based surveillance system known as CORONA. This system, in turn, provided the intelligence that in 1961 refuted the claims of a "missile gap" that had been so prominent in the 1960 U.S. presidential campaign.

According to Heppenheimer's analysis, there may be doubt therefore as to whether Sputnik really did demonstrate the USSR's superior scientific and technological prowess. There can however be no doubt that it was interpreted in this way by U.S. political leaders, the press, and public opinion. In this way, Sputnik served as a powerful catalyst for major new research and educational initiatives in science, mathematics, and engineering in the United States, beginning in the late 1950s and continuing at least into the 1970s.

APPENDIX B
Evolution of the National Institutes of Health

The history of the National Institutes of Health (NIH), now the world's largest and most eminent supporter of basic science, is itself of considerable interest to our story, although of course we cannot discuss it in any detail. The NIH began as a one-room Hygienic Laboratory created in 1887 within the Marine Hospital Service (MHS), which had been established to provide medical care for merchant seamen and later became the U.S. Public Health Service (PHS). This Hygienic Laboratory had only one full-time staff member until 1901; originally it was located on Staten Island in New York, but moved to Washington near the U.S. Capitol in 1891. In 1902, the Biologics Control Act charged the Hygienic Laboratory with regulatory rather than research functions, focused on the safety and efficacy of vaccines and antitoxins.[1] In 1912, the Hygienic Laboratory, now part of a newly named Public Health Service (PHS), was authorized to conduct research into noncontagious diseases and into the pollution of streams and lakes.

In 1930, the Ransdell Act changed the name of the Hygienic Laboratory to the "National Institute of Health" (note the singular form), and authorized it to provide fellowships for research into basic biological and medical problems. In 1937, the National Cancer Institute (NCI) was established by law as an entity separate from the NIH. In an important departure, the new NCI was authorized to award "extramural" grants to nonfederal scientists for research on cancer, something that the NIH itself was not able to do.

The 1944 Public Health Service Act incorporated the NCI under the NIH umbrella, and also extended to the rest of NIH the NCI's authorization to provide extramural research grants. From 1946 onward, the funds provided for such extramural grants began to grow rapidly. The total was about $4 million in 1947; ten years later it was more than $100 million. By 1974, it had reached $1 billion.

Also beginning at the end of World War II, the original "divisions" of NIH were transformed into new "Institutes." Initially these had academic medical names such as the National Microbiological Institute (NMI) and the Experimental Biology and Medicine Institute (EMBI), but in the 1950s the practice shifted to naming institutes after diseases out of the "conviction that institutes named after diseases stood a better chance for being funded by Congress." Meanwhile advocacy organizations for improved research and treatment of particular health problems successfully lobbied Congress to create additional institutes on heart disease, dental diseases, and mental health. As new institutes proliferated, the singular National Institute of Health was formally renamed in the plural form, the National Institutes of Health. By 1960, there were ten institutes and centers at the NIH; by 1970 there were fifteen; and by 1998 the NIH had twenty-seven institutes and centers.[2] Figures 2.5 and 2.6 provide timelines of NIH funding from 1962 to 2012, in both current and constant dollars.

APPENDIX C
"A Nation at Risk" and the Sandia Critique

In 1983, a commission appointed by Ronald Reagan's Secretary of Education Terrel Bell produced a report on the deterioration of U.S. education. The report carried the evocative title *A Nation at Risk*.[1] Its themes and rhetorical tone may be best appreciated by reading its opening two paragraphs:

> Our Nation is at risk. Our once unchallenged preeminence in commerce, industry, science, and technological innovation is being overtaken by competitors throughout the world ... the educational foundations of our society are presently being eroded by a rising tide of mediocrity that threatens our very future as a Nation and a people. What was unimaginable a generation ago has begun to occur—others are matching and surpassing our educational attainments.
>
> If an unfriendly foreign power had attempted to impose on America the mediocre educational performance that exists today, we might well have viewed it as an act of war. As it stands, we have allowed this to happen to ourselves ... We have, in effect, been committing an act of unthinking, unilateral educational disarmament.

In his 1988 memoir, Bell wrote that in the early 1980s he had been looking for what he called a "Sputnik-type occurrence" that would serve to dramatize all the "constant complaints about education and its effectiveness."[2] The 1983 release of the *Nation at Risk* report by the commission Bell appointed, while it failed to produce a repeat of the shock and near-panic of the "Sputnik-type occurrence" described elsewhere in this volume, did indeed attract high levels of attention in the press. According to one tally, the

Washington Post alone carried twenty-eight stories on the report during the month following.[3]

It is not possible here to provide more than a cursory discussion of *A Nation at Risk*. The report stimulated its own vast literature of publications that include both strong support and harsh criticism, much of which was politicized or reflected interest-group perspectives.[4] For example, a 1990 report by Sandia National Laboratory requested by then-secretary of energy James Watkins concluded that serious statistical errors underlay the conclusions of the *Nation at Risk* report. In particular, the core conclusion that public education had deteriorated during the 1960s and 1970s was based upon misinterpretation of declines in average SAT scores from 1963 to 1980. Sandia found that these declines in average SAT scores were caused by changes in the composition of the pool taking the SAT test, rather than by deterioration of U.S. schools. Whereas in the 1950s most test-takers had come from more elite secondary schools, during the 1960s and 1970s the test expanded dramatically to include larger proportions from student categories that performed less well. The Sandia analysts attempted to control for this distortion of the averages, and found that all the student subgroups it examined produced SAT scores over this period that actually had been stable or rising.

As the results of the Sandia critique of *A Nation at Risk* became available, a Washington tempest ensued. Some alleged that its findings were being suppressed,[5] while others argued that *A Nation at Risk* reflected real problems with U.S. education that were minimized by weaknesses in the Sandia National Laboratory's appraisal. In an extensive assessment of the Sandia analysis published in 1994, Lawrence Stedman concluded that

> the Sandia analysts deserve great credit for restoring a balance in the discussion of U.S. education. In broad outlines, their perspective is right—performance generally has been stable over the past 2 decades, there have been improvements in some areas, the United States is a world leader in high school and college graduation, and ethnic and racial groups need to be targeted for assistance because they continue to lag behind. Their data-driven analyses are a welcome counterpoint to the critics' insistent trumpeting of a decline in excellence.
>
> Like the school critics, however, the Sandia analysts were too quick to believe that the standing of the United States in the international assessment was low, but, like other revisionists, they were too quick to dismiss them as of little value. The U.S. position is far more mixed, and there are important lessons to be drawn from the assessments.[6]

Although *A Nation at Risk* was produced by a conservative Republican administration, prominent conservatives such as William Buckley mocked its recommendations as banal, and even the humorist Russell Baker got in the act by writing that one of the Report's most cited phrases, "a rising tide of mediocrity," would not be worth "more than a C in tenth grade English," and that overall "I'm giving them an A+ in mediocrity."[7]

Whatever one may think of the merits or weaknesses of the *Nation at Risk* report, its focus was clearly on the K-12 levels. However, its authors and supporters argued strongly, both explicitly and implicitly, that what they saw as the failing performance of K-12 education was closely related to deteriorating outcomes in advanced levels of higher education in science and mathematics, and that these in turn implied disturbing implications for U.S. national security and for economic performance.

What such arguments ignore, however, is the fact that those entering into advanced graduate education in STEM fields have always represented only small fractions of any student cohort, drawn disproportionately from the highest-performing quartiles. The argued worsening of *average* K-12 performance, whether they reflected true deterioration or were merely an artifact of the extension of SAT exams to lower-performing subsets of students, might have little or no impact upon such high-performing secondary school students most likely to pursue advanced STEM degrees.

NOTES

Introduction

1. The first globalization wave occurred a century earlier, beginning around 1880 and ending with a bang with the onset of World War I in 1914.
2. The "Y2K" episode involved massive short-term expenditures and hiring by governments and corporations alike to "fix" legacy computer code and update equipment, without which some warned of catastrophic breakdowns of computerized control and data systems at midnight of December 31, 1999, as the calendar turned over from the year 1999 to 2000. By one estimate in the computer industry press, some $134 billion was spent in the late 1990s in the United States to address the issues involved. See Robert L. Mitchell, "Y2K: The good, the bad and the crazy," *Computer World*, December 28, 2009.

   A compendium of materials on the Y2K episode, assembled by the International Y2K Cooperation Center (funded by the World Bank), is deposited at the Charles Babbage Institute, University of Minnesota. Available online at http://discover.lib.umn.edu/cgi/f/findaid/findaid-idx?c=umfa;cc=umfa;rgn=main;view=text;didno=cbi00153.

Chapter 1: Recent Alarms

1. "Tapping America's Potential: The Education for Innovation Initiative," Washington, DC: Business Roundtable, July 2005. Available online at http://www.uschamber.com/sites/default/files/reports/050727_tapstatement.pdf.
2. Committee on Prospering in the Global Economy of the 21st Century, National Research Council, *Rising Above the Gathering Storm: Energizing and Employing America for a Brighter Economic Future* (Washington, DC: National Academies Press, 2007). Available online at http://www.nap.edu/catalog.php?record_id=11463#toc.
3. Council on Competitiveness, *Innovate America: National Innovation Initiative Summit and Report* (Washington, DC: Council on Competitiveness, 2005), pp. 80–90. The author was a participant in one of the working groups. Available online at http://www.compete.org/images/uploads/File/PDF%20Files/NII_Innovate_America.pdf.
4. Council on Competitiveness, *Innovate America*, p. 49.
5. "BRT member companies comprise nearly a third of the total value of the U.S. stock market and invest more than $150 billion annually in research and development—nearly half of all private U.S. R&D spending." For further information; see http://businessroundtable.org/about-us/.
6. Emphasis in original.
7. "Tapping America's Potential," 2005, p. 7.
8. "Tapping America's Potential," 2005, pp. 5–8.
9. "Tapping America's Potential," 2005, pp. 10–13.
10. http://www.tap2015.org/.
11. The National Research Council is the executive arm of the National Academies, the National Academy of Science, National Academy of Engineering, and Institute of Medicine.

12. Senators Lamar Alexander (R-TN); Senator Jeff Bingamen (D-NM); Rep. Sherwood Boehlert (R-NY); and Rep. Bart Gordon (D-TN).

13. These included current or former CEOs of Lockheed Martin, Intel, DuPont, Exxon Mobil, and Merck.

14. These included current or former presidents of MIT, University of Maryland, College Park, Yale University, Rensselaer Polytechnic Institute, and Texas A&M University.

15. Steven Chu, Joshua Lederberg, Robert C. Richardson.

16. *Gathering Storm*, p. xi.

17. Full disclosure: the author was one of these thirty-seven external reviewers.

18. The full list of the Principal Project Staff of the NRC report is available at: http://www .nap.edu/openbook.php?record_id=11463&page=R5.

19. *Gathering Storm*, pp. 23–25.

20. *Gathering Storm*, p. 94.

21. *Gathering Storm*, p. 31.

22. *Gathering Storm*, p. 104.

23. *Gathering Storm*, p. 94.

24. *Gathering Storm*, pp. 98–99.

25. *Gathering Storm*, p. 104.

26. The America Creating Opportunities to Meaningfully Promote Excellence in Technology, Education, and Science Act (COMPETES), H.R. 2272, S. 761, Conference completed July 31, 2007, Signed into Law by the President, August 9, 2007.

27. Committee on Science and Technology, U.S. House of Representatives, Legislative Highlights, July 31, 2007. The America Creating Opportunities to Meaningfully Promote Excellence in Technology, Education, and Science Act (COMPETES). Available online at http://science.house.gov/legislation/leg_highlights_detail.aspx?NewsID=1938.

28. Two of those no longer available had become members of the Cabinet: Robert Gates, Secretary of the Department of Defense, and Stephen Chu, Secretary of the Department of Energy.

29. Members of the 2005 "Rising Above the Gathering Storm" Committee, *Rising Above the Gathering Storm, Revisited: Rapidly Approaching Category 5*. Prepared for the Presidents of the National Academy of Sciences, National Academy of Engineering, and Institute of Medicine (Washington, DC: National Academies Press, 2010). Available online at http:// www.nap.edu/catalog.php?record_id=12999.

30. National Hurricane Center, National Weather Service, National Oceanic and Atmospheric Administration, "Saffir-Simpson Hurricane Wind Scale." A Category 5 hurricane is characterized by sustained winds in excess of 157 miles per hour (252 km/hr). As to types of damage: "Catastrophic damage will occur: A high percentage of framed homes will be destroyed, with total roof failure and wall collapse. Fallen trees and power poles will isolate residential areas. Power outages will last for weeks to possibly months. Most of the area will be uninhabitable for weeks or months." Available online at http://www .nhc.noaa.gov/sshws_table.shtml?large. Accessed August 17, 2013.

31. Hurricane Camille, another Category 5 storm in 1969, led to 256 deaths and $9.8 billion in damage. Hurricane Katrina, though a Category 3 storm when it made landfall in Louisiana in 2005, had previously reached Category 5 while over the Gulf of Mexico. Because its high storm surge hit populated and low-lying areas including the city of New Orleans, and because of inadequate preparations by federal, state, and local governments, it caused the deaths of approximately 1,500 people and damage estimated at nearly $85 billion. (All damage estimates adjusted for inflation to 2006 dollars.) For more information on the destructive effects of Category 5 storms, see Eric S. Blake, Edward N. Rappaport, and Christopher W. Landsea, "The Deadliest, Costliest, and Most Intense United States Tropical Cyclones from 1851 to 2006 (and other frequently requested hurricane

facts)." National Hurricane Center. National Oceanic and Atmospheric Administration, April 15, 2007. See especially tables 3a and 3b. Available online at http://www.nhc.noaa .gov/pdf/NWS-TPC-5.pdf.

32. B. Lindsay Lowell and Hal Salzman, "Into the Eye of the Storm: Assessing the Evidence on Science and Engineering Education, Quality, and Workforce Demand," Urban Institute, October 2007. Available online at http://www.urban.org/UploadedPDF/411562 _Salzman_Science.pdf. This research was supported based on a proposal submitted by the authors to the Alfred P. Sloan Foundation.

33. Lowell and Salzman, "Into the Eye of the Storm," p. ii.

34. Lowell and Salzman, "Into the Eye of the Storm," p. ii.

35. Lowell and Salzman, "Into the Eye of the Storm," p. ii. Available online at http://www .urban.org/UploadedPDF/411562_Salzman_Science.pdf.

36. Yu Xie and Alexandra A. Killewald, *Is American Science in Decline?* (Cambridge, MA and London: Harvard University Press, 2012), esp. pp. 30–34.

37. Xie and Killewald, *Is American Science in Decline?*

38. Xie and Killewald, *Is American Science in Decline?* pp. 116–19. The 3:1 ratio is reported by Lowell and Salzman, and also by the National Science Board, *Science and Engineering Indicators 2010*, pp. 3–6.

39. TIMSS (Trends in International Mathematics and Science Study) is a collaborative comparative study of mathematics and science achievement at (approximately) grades 4 and 8, initiated in 1995 and undertaken every four years with coordination by Boston College. More information is available at http://timssandpirls.bc.edu/timss2011/index.html. PISA (Programme for International Student Assessment) is an international study operated by the Organisation for Economic Co-operation and Development (OECD) in Paris. It seeks to evaluate education systems worldwide by testing the skills and knowledge of a sample of 15-year-old students. More information is available at http://www.oecd.org/ pisa/aboutpisa/ . These two internationally comparative studies have substantial overlaps but also differ in significant ways from one another in their design and implementation. See Xie and Killewald, *Is American Science in Decline?* pp. 30–31.

40. Xie and Killewald, *Is American Science in Decline?* pp. 36–37.

41. Similarly evocative language such as "a nation at risk" and "unilateral disarmament" has been used in both earlier and later reports. See discussion in chapter 2.

Chapter 2: No Shortage of Shortages

1. Or more accurately, the plural form "systems," given the high degree of local financing and control of K-12 education in the United States.

2. Hunter R. Rawlings, "Why Research Universities Must Change," *Inside Higher Education*, March 30, 2012. Available online at http://www.insidehighered.com/views/2012/03/30/ essay-research-universities-must-pay-more-attention-student-learning#ixzz1zgMngN6J.

3. The National Commission on Excellence in Education, *A Nation At Risk: The Imperative for Reform*, Report to the Nation and the Secretary of Education, United States Department of Education, April 1983, n.p.; quoted text from first page of report. Available online at http://www2.ed.gov/pubs/NatAtRisk/index.html.

4. David Kaiser, *American Physics and the Cold War Bubble* (Chicago: University of Chicago Press, forthcoming), ch. 1.

5. Kaiser, *American Physics and the Cold War Bubble*, ch. 1.

6. Kaiser, *American Physics and the Cold War Bubble*, introduction.

7. Kaiser, *American Physics and the Cold War Bubble*, ch. 1.

8. William L. O'Neill, *American High: The Years of Confidence, 1945–1960* (New York: Free Press, 1989), p. 136.

9. Nicholas DeWitt, *Soviet Professional Manpower: Its Education, Training, and Supply* (Washington, DC: National Science Foundation, 1955).

10. Alexander Korol, *Soviet Education for Science and Technology* (Cambridge, MA: MIT Press, 1957). Citation at pp. xi, 391, 400, 407–8 ("unwarranted implications"), 414.

11. David Kaiser, "The Physics of Spin: Sputnik Politics and American Physicists in the 1950s," *Social Research* 73, no. 4 (Winter 2006), pp. 1228–31. See also Kaiser, *American Physics and the Cold War Bubble*.

12. Kaiser, "The Physics of Spin," p. 1231.

13. Lee DuBridge, Testimony before National Committee for the Development of Scientists and Engineers, June 1956, quoted in National Science Foundation, *Sixth Annual Report, 1956*, p. 13. Available online at http://www.nsf.gov/pubs/1956/annualreports/ar_1956_sec3.pdf. Also quoted in Kaiser, "The Physics of Spin," p. 1232.

14. John Krige, "NATO and the Strengthening of Western Science in the post-Sputnik Era," *Minerva* 38 (2000), pp. 81–108, p. 88.

15. Kaiser, "The Physics of Spin," pp.1238–39.

16. Krige, "NATO and the Strengthening of Western Science," pp. 89–90.

17. At that time NATO was an alliance of 15 member states—13 in Western Europe plus the United States and Canada. There are now 28 members of NATO, 26 of which are in Europe.

18. Krige, "NATO and the Strengthening of Western Science," p. 91.

19. Jackson report as cited in Kaiser, *American Physics and the Cold War Bubble*.

20. The IGY involved 67 countries including the United States and the USSR. The "Year" was actually an 18-month program scheduled from July 1, 1957 to December 31, 1958 to take advantage of high levels of solar activity.

21. Now the International Council for Science, headquartered in Paris. The organization was founded in 1931, and is perhaps the leading nongovernmental organization devoted to international cooperation in the advancement of science. Its current members include national scientific bodies (120 members, representing 140 countries) and International Scientific Unions (31 members). Available online at http://www.icsu.org/about-icsu/about-us.

22. T. A. Heppenheimer, *Countdown: A History of Flight* (New York: John Wiley and Sons, 1997), pp. 92–95.

23. Heppenheimer, *Countdown*, pp. 94–95.

24. Heppenheimer, *Countdown*, pp. 95–96.

25. Cited in Heppenheimer, *Countdown*, p. 97.

26. Heppenheimer, *Countdown*, p. 91.

27. Heppenheimer, *Countdown*, p. 91.

28. Heppenheimer, *Countdown*, p. 100. The army's launch vehicle, the Jupiter-C, was on its launch pad on September 20, 1956, though not yet equipped with the final stage 4 needed to launch a satellite into orbit, when it was put on hold. See appendix A.

29. Walter McDougall, *The Heavens and the Earth: A Political History of the Space Age* (New York: Basic Books, 1985), pp. 123–24.

30. Indeed, some political insiders already had commented on the importance of the United States being the first to launch an Earth-orbiting satellite: Nelson Rockefeller, then an assistant in the Eisenhower White House, had responded to a report by the assistant secretary of Defense for Research and Development that "I am impressed by the costly consequences of allowing the Russian initiative to outrun ours through an achievement that will symbolize scientific and technological advancement to people everywhere. The

stake of prestige involved makes this a race that we cannot afford to lose." Nelson
Rockefeller, 1955, quoted in Heppenheimer, *Countdown*, pp. 99–100.

31. Roger D. Launius, "Sputnik and the Origins of the Space Age" (Washington, DC:
National Aeronautics and Space Administration), n.d. Available online at http://history
.nasa.gov/sputnik/sputorig.html.

32. Kaiser, "The Physics of Spin," p. 1233.

33. Launius, "Sputnik and the Origins of the Space Age."

34. Launius, "Sputnik and the Origins of the Space Age."

35. T. A. Heppenheimer, "How America Chose Not to Beat Sputnik into Space," *Invention and
Technology* (Winter 2004), p. 48.

36. Launius, "Sputnik and the Origins of the Space Age."

37. Public Law 85-325.

38. Richard Van Atta, *Fifty Years of Innovation and Discovery*. Available online at http://www
.arpa.mil/Docs/Intro_-_Van_Atta_200807180920581.pdf.

39. John Gunther, *Inside Russia Today* (New York: Harper, 1958), pp. 257–58.

40. Kaiser, *American Physics and the Cold War Bubble.*

41. Pamela Ebert Flattau, Jerome Bracken, Richard Van Atta, Ayeh Bandeh-Ahmadi, Rodolfo
de la Cruz, Kay Sullivan, *The National Defense Education Act of 1958: Selected Outcomes*,
IDA Document D-3306 (Washington, DC: Science and Technology Policy Institute,
Institute for Defense Analyses, March 2006), pp. I-1 to I-2 and VII-1. Available online at
http://www.ida.org/stpi/pages/D3306-FINAL.pdf.

42. Christopher A. Preble, "Who Ever Believed in the 'Missile Gap'? John F. Kennedy and the
Politics of National Security." *Presidential Studies Quarterly* 33, no. 4 (December 2003),
801–26, pp. 814–25.

43. President John F. Kennedy, "Special Message to the Congress on Urgent National Needs,"
Delivered in person before a Joint Session of Congress, May 25, 1961, p. 4. Historical
Resources, John F. Kennedy Presidential Library and Museum, Boston, MA. Available
online at http://www.jfklibrary.org/Historical+Resources/Archives/Reference+Desk/
Speeches/JFK/Urgent+National+Needs+Page+4.htm.

44. Charles D. Benson and William Barnaby Faherty. "Mid-1963: A Time of Reappraisal," in
Moonport: A History of Apollo Launch Facilities and Operations, NASA Special Publication-
4204 in the NASA History Series, 1978, ch. 7, p. 6. Available online at http://www.hq.nasa
.gov/office/pao/History/SP-4204/ch7-6.html.

45. Benson and Faherty, "Mid-1963," ch. 7, p. 6.

46. Benson and Faherty, "Mid-1964," ch. 7, p. 6.

47. *Washington Post*, September 21, 1963, p. A-10. Cited in Benson and Faherty, "Mid-1963,"
ch. 7, p. 6.

48. NASA History Office, "Introduction," *Apollo 30th Anniversary*, updated September 20,
2001. Available online at http://history.nasa.gov/ap11ann/introduction.htm.

49. http://history.nasa.gov/ap11ann/top10sci.htm.

50. NASA Langley Research Center, "NASA Langley Research Center's Contributions to the
Apollo Program." Available online at http://www.nasa.gov/centers/langley/news/
factsheets/Apollo.html. Detailed cost accounting is provided by NASA headquarters at
http://www.hq.nasa.gov/office/pao/History/SP-4214/app2.html.

51. A second accounting of appropriations is summarized in Richard W. Orloff, *Apollo by the
Numbers: A Statistical Reference* (NASA History Series, NASA SP-2000-4029). Available
online at http://history.nasa.gov/SP-4029/Apollo_18-16_Apollo_Program_Budget
_Appropriations.htm.

52. http://www.measuringworth.com/uscompare/?redirurl=calculators/uscompare/.

53. http://history.nasa.gov/SP-4029/Apollo_18-16_Apollo_Program_Budget_Appropriations
.htm.

54. I am indebted to David Kaiser for drawing my attention to the Project Hindsight study. Office of the Director of Defense Research and Engineering, Project Hindsight, Final Report (Washington, DC: October 1969), 214 pp. Available online at http://www.dtic .mil/cgi-bin/GetTRDoc?Location=U2&doc=GetTRDoc.pdf&AD=AD0495905.
55. Project Hindsight, 1969, p. xiv.
56. Project Hindsight, 1969, p. xvii.
57. Project Hindsight, 1969, p. xxiii.
58. Project Hindsight, 1969, pp xvi–xvii.
59. James E. Westheider, *The Vietnam War* (Santa Barbara, CA: Greenwood Press, 2007), pp. 33–35.
60. Joseph P. Martino, "Science and Technology for Economic Ends," in *The Academy in Crisis: The Political Economy of Higher Education*, 2nd edition, John Sommer, ed., pp. 257–86 (New Brunswick, NJ: Transaction Publishers; 1994). Citation at p. 279. See also Stuart Umpleby, "Heinz von Foerster and the Mansfield Amendment," *Cybernetics and Human Knowing* 10, nos. 3–4, (December 2002), pp. 187–90.
61. Martino, "Science and Technology for Economic Ends," p. 279.
62. According to the National Bureau of Economic Research (NBER), the official arbiter of U.S. economic cycles, peak economic activity occurred in December 1969, followed by an eleven-month economic decline that bottomed out around November 1970. Although the GDP decline was only about 0.6 percent and unemployment rose only to 6.1 percent, the recession of 1969–70 took place after an unusually long period of economic expansion: the U.S. economy had been expanding for fully 106 months from the previous economic trough in February 1961, and hence the 1960s had seen what was then by far the longest economic expansion since 1854. See National Bureau of Economic Research, *US Business Cycle Expansions and Contractions* (Cambridge, MA: NBER). Available online at http://www.nber.org/cycles/. Accessed August 7, 2013.
63. See, for example, Michel Morange, *A History of Molecular Biology* (Cambridge, MA: Harvard University Press, 1998).
64. National Science Board, *Science and Engineering Indicators 2012*, table 5-6, p. 5-21 and fig. 5-14, p. 5-23.
65. R. A. Alberty, "Projection of Doctoral Chemists and Physicists to 1980" (MIT Manpower Symposium, Cambridge, MA, May 1970), cited in Wallace R. Brode, "Manpower in Science and Engineering, Based on a Saturation Model," *Science*, New Series, 173, no. 3993 (July 16, 1971), pp. 206–13. (Stable URL: http://www.jstor.org/stable/1732663), citation on p. 212.
66. Brode had been elected president of the American Chemical Society, the Optical Society of America, and the American Association for the Advancement of Science. He also was the 1960 recipient of the Priestley Medal, the highest honor conferred by the American Chemical Society.
67. Brode, "Manpower in Science and Engineering," citation on p. 206.
68. Brode, "Manpower in Science and Engineering," p. 209.
69. Brode, "Manpower in Science and Engineering," p. 206.
70. Brode, "Manpower in Science and Engineering," pp. 208–211. Emphasis added.
71. National Bureau of Economic Research, *US Business Cycle Expansions and Contractions* (Cambridge, MA: NBER). Available online at http://www.nber.org/cycles/. Accessed August 7, 2013.
72. Office of the President, *Budget of the United States Government: Historical Tables Fiscal Year 2008*, Table 8.2—Outlays by Budget Enforcement Act Category in Constant (FY2000) Dollars: 1962–2012. Available online at http://www.gpoaccess.gov/USbudget/fy08/hist.html.

73. In the following year, Bloch and his retired IBM colleagues Frederick P. Brooks and Bob O. Evans received the National Technology Medal from U.S. President Ronald Reagan for their contributions to the development of the IBM System/360.

74. Statement of Mr. Erich Bloch, Director, National Science Foundation, on National Science Foundation Authorization Act for Fiscal Year 1987, Hearing before the Committee on Labor and Human Resources, U.S. Senate, March 26, 1986. Reproduced in full in *Hearing before the Subcommittee on Investigations and Oversight, Committee on Science, Space, and Technology, U.S. House of Representatives, One Hundred Second Congress*, April 8, 1992, pp. 380–401.

75. Steven Weinberg, "The Crisis of Big Science," *New York Review of Books*, May 10, 2012. Available online at http://www.nybooks.com/articles/archives/2012/may/10/crisis-big -science/?pagination=false&printpage=true }.

76. http://www.tcf.ua.edu/AZ/ITHistoryOutline.htm.

77. Barry M. Leiner, Vinton G. Cerf, David D. Clark, Robert E. Kahn, Leonard Kleinrock, Daniel C. Lynch, Jon Postel, Larry G. Roberts, Stephen Wolff, *A Brief History of the Internet*, The Internet Society, 2010. Available online at http://www.isoc.org/internet/ history/brief.shtml#Commercialization.

78. U.S. Department of Homeland Security, U.S. Citizenship and Immigration Services, "H-1B Specialty Occupations, DOD Cooperative Research and Development Project Workers, and Fashion Models, Eligibility Criteria." Available online at http://www.uscis.gov/.

79. http://www.govtrack.us/congress/billtext.xpd?bill=s106-2045.

80. Presentation to workshop on U.S. scientific workforce by Kevin Casey, Associate Vice President for the Office of Government, Community, and Public Affairs, Harvard University, National Bureau of Economic Research, Cambridge MA, January 24, 2002. Available online at http://www.nber.org/~sewp/events/2002.01.24/2002jan24AGEN.pdf.

81. Leon M. Lederman, *Science: The End of the Frontier?* Supplement to *Science*, 251 no. 4990 (January 11, 1991), 21pp. The informal survey was sent to the chairs of the physics, chemistry, and biology departments of 50 universities—the 30 largest research universities plus 20 others covering a range of less research-intensive institutions. Each chair was asked to respond personally to the informal survey letter, and also to forward to a few faculty members including the youngest full professor and at least one third- or fourth-year assistant professor in the department. About 250 responses were received.

82. Lederman, *Science: The End of the Frontier?* p. 6.

83. Lederman, *Science: The End of the Frontier?* p. 12.

84. Lederman, *Science: The End of the Frontier?* p. 17.

85. Lederman, *Science: The End of the Frontier?* p. 13.

86. Lederman, *Science: The End of the Frontier?* p. 18.

87. In full disclosure, the author is a Fellow of the AAAS, and has in the past served as the elected chair of its Section on the Social and Behavioral Sciences.

88. Joseph Palca, "Leon Lederman's Quest: Double Science Funding," *Science* 251 (January 11, 1991), pp. 153–54.

89. Examples include the American Cancer Society, American Heart Association, American Lung Association, Alzheimer's Association, American Diabetes Association. For a full list of the 500 current members and supporters of Research!America, see http://www .researchamerica.org/members_supporters#6.

90. Jeffrey Mervis, "Research Funding: Senate Bills Back Huge Increases," *Science* 275, no. 5300 (January 31, 1997), pp. 608.

91. Mervis, "Research Funding," p. 608.

92. Andrew Lawler, "The 1999 Budget: Science Catches Clinton's Eye," *Science* 279, no. 5352 (February 6, 1998), p. 794

93. Eliot Marshall, "The 1999 Budget: Science Funding: Up in Smoke?" *Science* 279, no. 5353 (February 13, 1998), pp. 974–75.
94. Marshall, "The 1999 Budget: Science Funding: Up in Smoke?"
95. David Malakoff and Eliot Marshall, "1999 Budget Finale: NIH Wins Big as Congress Lumps Together Eight Bills," *Science* 282, no. 5389 (October 23, 1998), p. 598.
96. David Malakoff, "Budget Finale: NIH Gets $17.9 Billion in Another Record Year," *Science* 286, no. 5445 (November 26, 1999), p. 1654.
97. "Double Trouble," *Science* 289, no. 5478 (July 21, 2000), p. 373.
98. "Gore and Bush Offer Their Views on Science," Policy Forum, *Science* 290, no. 5490 (October 13, 2000), pp. 262–69.
99. David Malakoff, "2001 Spending: NIH Gets $2.5 Billion More as Congress Wraps Up Budget," *Science* 290, no. 5500 (December 22, 2000), p. 2226.
100. David Malakoff, "Lobbying: Can ASTRA Restore a Glow to the Physical Sciences?" *Science* 292, no. 5518 (May 4, 2001), p. 832.
101. David Malakoff, "U.S. Budget: Biomedicine Gets Record Raise as Congress Sets 2002 Spending," *Science* 295, no. 5552 (January 4, 2002), pp. 24–25.
102. David Malakoff, "2003 Budget: Science Agencies Get Most of What They Want, Finally" *Science* 299, no. 5610 (February 21, 2003), pp. 1160–61.
103. The FY 2008 budget for NIH was $29.457 billion, up by about $2.2 billion and 8 percent from the 2003 level. Available online at http://officeofbudget.od.nih.gov/ui/2008/Summary%20of%20FY%202009%20Budget-Press%20Release.pdf.
104. National Institutes of Health, *NIH External Data Book 2007*, Research Project Grants. Available online at http://grants2.nih.gov/grants/award/Research_Training_Investment/Research_Training_Investment.cfm.
105. National Institutes of Health, Office of External Research, presentation to Association of Independent Research Institutes, Phoenix, AZ, October 3, 2006, fig. 25. Available online at http://grants.nih.gov/grants/award/trends/airi2006.ppt.
106. The GDP Chain Index is a standard measure of price increases in the broad economy produced by the Bureau of Economic Analysis, US Department of Commerce. Available online at http://officeofbudget.od.nih.gov/gbiPriceIndexes.htm.
107. The BRDPI index is designed to reflect price increases for biomedical R&D that typically are higher than those for the broad economy. The index is produced by the Bureau of Economic Analysis under an interagency agreement with the NIH, and has acknowledged imperfections, e.g. it adjusts for changes in the price and cost of research inputs, but not for the cost of research "output." Available online at http://officeofbudget.od.nih.gov/gbiPriceIndexes.html.
108. See http://officeofbudget.od.nih.gov/UI/GDP_FromGenBudget.htm.
109. S. J. Heinig, J. Y.Krakower, H. B Dickler, and D. Korn, "Sustaining the Engine of U.S. Biomedical Discovery, *New England Journal of Medicine* 357, 10 (September 6, 2007), fig. 1.
110. D. Korn, R. R. Rich, H. H. Garrison, S. H. Golub, M.J.C. Hencrix, S. J. Heinig, B. S. Masters, and R. J. Turman, "The NIH Budget in the "Postdoubling" Era," *Science*, 296, no. 1401 (May 24, 2002).
111. R. Freeman and J. Van Reenen, "What if Congress Doubled R&D Spending on the Physical Sciences?" Presented at the NBER Conference on *Innovation Policy and the Economy*, Washington, DC, April 15, 2008.
112. http://report.nih.gov/NIH_Investment/PPT_sectionwise/NIH_Extramural_Data_Book/RPG.ppt.
113. National Institutes of Health, *NIH External Data Book 2007*, Research Project Grants. Available online at http://grants2.nih.gov/grants/award/Research_Training_Investment/Research_Training_Investment.cfm.

114. "A Broken Pipeline? Flat Funding of the NIH Puts a Generation of Science at Risk." Report by seven research institutions released March 11, 2008. Available online at http://www.brokenpipeline.org/.

115. http://www.brokenpipeline.org/.

116. Andy Guess, "Clarion Call for More NIH Funding," *Inside Higher Ed*, March 12, 2008. Available online at http://www.insidehighered.com/news/2008/03/12/nih.

117. National Research Council, Office of Scientific and Engineering Personnel, *Forecasting Demand and Supply of Doctoral Scientists and Engineers: Report of a Workshop on Methodology* (Washington, DC: National Academy Press, 2000), p. 1.

Chapter 3: Beliefs, Interests, Effect

1. See, for example, Thomas C. Reed and Danny B. Stillman, *The Nuclear Express: A Political History of the Bomb and its Proliferation* (Minneapolis, MN: Zenith Press, 2009.); and Richard Rhodes, *Dark Sun: The Making of the Hydrogen Bomb* (New York: Simon & Schuster, 1995).

2. So too would any consumer needing to hire a plumber to fix a domestic water leak, though few would claim that plumbers' high hourly wages were due to an inherent numerical "shortage" of people capable of such plumbing work.

3. Data are from Bureau of Labor Statistics, "Public Data: Unemployment rate, not seasonally adjusted, Fresno Country, CA," as presented in Google, "Public Data," available online at ashttp://www.google.com/publicdata/explore?ds=z1ebjpgk2654c1_&met _y=unemployment_rate&idim=county:CN060190&fdim_y=seasonality:U&dl=en&hl =en&q=unemployment+rate+in+fresno+county,+ca#!ctype=l&strail=false&bcs=d&nselm =h&met_y=unemployment_rate&fdim_y=seasonality:U&scale_y=lin&ind_y=false &rdim=country&idim=county:CN060190&ifdim=country&hl=en_US&dl=en&ind =false. Accessed August 9, 2013.

4. J. J. Dooley, "The Rise and Decline of U.S. Private Sector Investments in Energy R&D since the Arab Oil Embargo of 1973." Report prepared for the U.S. Department of Energy under Contract DE-AC05-76RL01830. Richland, WA, Pacific Northwest National Laboratory, November 2010 (PNNL-19958), pp. 5–7.

5. International Atomic Energy Agency, IAEA, "50 Years of Nuclear Energy," n.d., p. 4. Available online at http://www.iaea.org/About/Policy/GC/GC48/Documents/gc48inf-4 _ftn3.pdf

6. For a full listing of major government contractors as of 2010, see "2010 Washington Technology Top 100 Government Contractors," *Washington Technology*. Available online at http://washingtontechnology.com/toplists/top-100-lists/2010.aspx?Sort=Company -Name.

Chapter 4: The Influence of Employer and Other Interest Groups

1. Patrick Thibodeau and Sharon Machlis, "The data shows: Top H-1B users are offshore outsourcers: U.S. government's H-1B visa list shows accelerating demand from off-shore outsourcers," *Computerworld*, February 13, 2013. Available online at http://www

.computerworld.com/s/article/print/9236732/The_data_shows_Top_H_1B_users_are
_offshore_outsourcers?taxonomyName=Gov%27t+Legislation%2FRegulation&
taxonomyId=70.

2. N. S. Ramnath, "How Cognizant Overtook Infosys," *Forbes India*, 08/07/2012. Available
online at http://forbesindia.com/blog/technology/how-cognizant-overtook-infosys/.
See also "Cognizant 2013 Corporate Fact Sheet," http://www.cognizant.com/Recent
Highlights/Corporate_Fact_Sheet.pdf. Accessed August 17, 2013.

3. Cognizant Technology Solutions Corporation, "Form 10-Q, Quarterly Report pursuant
to Section 13 or 15(d) of the Securities Exchange Act of 1934, For the quarterly period
ended September 30, 2008." United States Securities and Exchange Commission, Commission File Number 0-24429, pp. 28–29. This SEC Filing is available online at http://
ctsh.edgarpro.com/edgar_conv_html%5C2008%5C11%5C07%5C0001193125-08-229956
.html#D10Q_HTM_TX93203_11.

4. Thibodeau and Machlis, "The data shows."

5. Ron Hira, "Top 10 users of H-1B guest worker program are all offshore outsourcing
firms"; http://www.epi.org/blog/top-10-h1b-guestworker-offshore-outsourcing/. Posted
February 14, 2013. Hira is co-author of the influential 2005 book, *Outsourcing America:
What's Behind Our National Crisis and How We Can Reclaim American Jobs* (New York:
AMACOM Books, 2005).

6. EB-2 refers to the "Employment-based Second Preference" visa for permanent immigration; EB-3 to the "Employment-based Third Preference" visa. For more information, see
U.S. Citizenship and Immigration Services, "Permanent Workers," Second Preference
EB-2 and Third Preference EB-3; available online at http://www.uscis.gov/portal/site/
uscis/ under "Working in the US," "Permanent Workers."

7. Information Technology Association of America, *Help Wanted: The IT Workforce at the
Dawn of a New Century* (Arlington, VA: Information Technology Association of America,
1997).

8. See biosketch on Anderson at http://www.nfap.com/about/biographies/.

9. In 1997 Anderson left the Cato Institute to become a staff member and ultimately staff
director of the Senate Immigration Subcommittee, then chaired by Senator Spencer
Abraham (R-MI), a longtime Cato associate. In these roles he reportedly was active in
blocking implementation of a new student visa system (see discussion later in this
chapter), and in staff efforts to pass the 2000 legislation that tripled the number of H-1B
visas each year. After Abraham lost his reelection bid in November 2000 and later was
appointed secretary of Energy in the incoming Bush administration, Anderson was
appointed executive associate commissioner for Policy and Planning and counselor to
the Commissioner in the Immigration and Naturalization Service. He departed the
Bush administration in 2003 and founded a new right-libertarian think tank focused on
immigration with the impressive title of the "National Foundation for American Policy,"
of which he is executive director and for which Senator Abraham is one of four listed
advisors. Available online at http://www.nfap.com/about/biographies/.

10. Personal communication, Harris Miller, president, Information Technology Association
of America, New York, 1998.

11. U.S. Department of Commerce, assistant secretary for Technology Policy, Office of
Technology Policy, "America's New Deficit: The Shortage of Information Technology
Workers" (Washington, DC: U.S. Department of Commerce, Fall 1997), 38pp; citation
on p. 7.

12. Information Technology Association of America, *Help Wanted: A Call for Collaborative
Action for the New Millennium* (Arlington, VA: Information Technology Association of
America, 1998). This report was financed and published by ITAA and written at Virginia
Polytechnic Institute and State University. Its estimate of 346,000 claimed vacancies for IT

was based upon a random sampling of 1,500 IT and non-IT companies with 100 or more employees. Its response rate was 36 percent (specifically, of 1,493 telephone interviews attempted, 532 were successfully completed).

13. Later renamed the Government Accountability Office, with the same initials.

14. United States General Accounting Office (GAO), *Information Technology: Assessment of the Department of Commerce's Report on Workforce Demand and Supply*, Report to Congressional Requesters, March 1998, 18pp. GAO/HEHS-98-106. Available online at http://www.gao.gov/assets/230/225415.pdf.

15. Information Technology Association of America, *Bridging the Gap: Information Technology Skills for a New Millennium* (Arlington, VA: ITAA, April 2000).

16. In 2008, ITAA merged with the Cyber Security Industry Alliance (CSIA) and the Government Electronics Industry Association (GEIA) and these then merged with the American Electronics Association (AeA) to form a new trade association conglomerate named TechAmerica. ITAA's then-president and CEO Phillip J. Bond became president of TechAmerica, which claims to represent nearly 1,500 companies. Bond, a former executive at Hewlett-Packard Corporation, was also a congressional staff member during the 1990s, and from 2001 to 2005 was Under Secretary of the U.S. Department of Commerce for Technology—in which capacity he may have overseen the operations of agencies such as the Office of Technology Policy, the group that produced the 1998 report based on the ITAA reports that had attracted such heavy methodological criticism in the 1998 GAO assessment cited earlier. See bio of Phillip J. Bond at www.techamerica.org/philbond.

17. For more information on ITI, see http://www.itic.org/.

18. Jim Snyder, "Oracle's Hoffman to Cognizant," *The Hill*, September 8, 2009. Available online at http://thehill.com/business-a-lobbying/57763-oracles-hoffman-to-cognizant.

19. See, for example, Rachel Burstein, "Overseas Invasion," *Mother Jones* (January/February 1998). Available online at http://motherjones.com/politics/1998/01/overseas-invasion.

20. Usefully available in a searchable database at http://elenasinbox.com.

21. Patrick Thibodeau, "'Elena's Inbox' details H-1B battle in Clinton White House: Memos to Supreme Court nominee Elena Kagan from Clinton administration opens door to battle over H-1B visa in critical year," *Computerworld*, July 2, 2010. Available online at http://www.computerworld.com/s/article/9178806/_Elena_s_Inbox_details_H_1B _battle_in_Clinton_White_House.

22. See bio on Julie A. Fernandes at http://www.americanprogressaction.org/events/2008/inf/ FernandesJulie.html.

23. HIB update email from Julie A. Fernandes to Bruce N. Reed and Elena Kagan, 10:07 am 01 May 1998, available online at http://elenasinbox.com/thread/hib-update-2/?q=Intel. See also "LRM #IMS 309 - H1B Temporary Immigrant Visa Program Reforms," email of 6:22 pm 27 Apr 1998. Available online at http://elenasinbox.com/thread/lrm-ims-309 -h1b-temporary-immigrant-visa-program-r/?q=$75,000.

24. According to its website, the Sunlight Foundation is "a nonpartisan nonprofit founded in 2006 that uses the power of the Internet to catalyze greater government openness and transparency. We do so by creating tools, open data, policy recommendations, journalism and grant opportunities to dramatically expand access to vital government information to create accountability of our public officials. Our vision is to use technology to enable more complete, equitable and effective democratic participation. Our overarching goal is to achieve changes in the law to require real-time, online transparency for all government information, with a special focus on the political money flow and who tries to influence government and how government responds." The Sunlight Foundation's principal funders include the Omidyar Network (established by the founder of eBay); Mike Klein (a wealthy lawyer and businessman who co-founded the Foundation); the William and Flora Hewlett Foundation; the Ford Foundation; and the Open Society Foundations

(founded and chaired by George Soros, the billionaire investor and philanthropist of progressive causes). For additional information, see http://www.sunlightfoundation.com/about/.

25. This searchable database is available at www.sunlightfoundation.org.
26. Anapama Narayanswamy, "Immigration: Give me your poor, your tired ... your lobbyists?" Sunlight Foundation blog, March 21, 2013. Available online at http://reporting.sunlightfoundation.com/2013/immigration/. Accessed July 4, 2013.
27. U.S. House of Representatives, Office of the Clerk, "Lobbying Disclosure Guidance," effective January 1, 2008, last revised February 15, 2013, reviewed June 18, 2013. Available online at http://lobbyingdisclosure.house.gov/amended_lda_guide.html. Accessed August 2, 2013.
28. Narayanswamy, "Immigration: Give me your poor, your tired ... your lobbyists?"
29. Investigations of the Facebook initial public offering were initiated by multiple regulators from federal and state governments and from within the financial industry itself. Some 40 lawsuits, including a $2.5 billion class action, were filed against the company, its lead investment banker Morgan Stanley, and the NASDAQ stock exchange; some of these have been settled, while others continue to wend their way slowly through the legal system. For a brief summary, see Sam Gustin, "Facebook's IPO One Year Later: Mobile Growth, Legal Headaches, and a Stalled Stock Price," *Time*, May 17, 2013. Available online at http://business.time.com/2013/05/17/facebooks-ipo-one-year-later-mobile-growth-legal-headaches-and-a-stalled-stock-price/#ixzz2YJukDILA. Accessed July 6, 2013.
30. Complete lists of FWD.US founders and major contributors can be found at http://www.fwd.us/our_supporters. Accessed July 5, 2013.
31. Narayanswamy, "Immigration: Give me your poor, your tired ... your lobbyists?
32. http://soprweb.senate.gov/index.cfm?event=getFilingDetails&filingID=D9ADD2DE-5409-43E3-AA87-ED459A95B96B&filingTypeID=51. Accessed July 5, 2013.
33. Giovanni Facchini, Anna Maria Mayda, and Prachi Mishra, "Do Interest Groups Affect U.S. Immigration Policy?" IMF Working Paper WP/08/244, October 2008, p. 6. The authors attempted but were unable to obtain comparable lobbying expenditure data by groups opposed to expanded immigration. In the absence of such information, they attempted to analyze the impacts of such groups using very different indicators, based upon assumptions that (1) lobbying in opposition to expanded immigration in the 2001–2005 period was dominated by organized labor and (2) that the size of union membership would provide a proxy measure of union lobbying expenditures. The union membership rate data across sectors were drawn from the Current Population Survey. On the basis of these assumptions, the authors reported that "a 1 percentage point increase in the union membership rate is associated with a 2.6–5.5 percent lower number of visas." Unfortunately both assumptions were rather weakly founded—opposition to expanded immigration during this period was hardly limited to organized labor, and while the size of union membership is a quantity that is available, it is likely only loosely related, if at all, to a union's direct lobbying expenditures on a particular type of legislation. Hence these estimates lack the specificity of the study's findings on business lobbying expenditures.
34. Transcript of Oral Testimony by Bill Gates, Chairman, Microsoft Corporation, before the United States Senate Committee on Health, Education, Labor, and Pensions, "Strengthening American Competitiveness for the 21st Century," Washington, DC, March 7, 2007 Available online at http://www.microsoft.com/Presspass/exec/billg/speeches/2007/03-07Senate.mspx.
35. Published on Computerworld Blogs at http://blogs.computerworld.com.
36. TechCrunch is a visible blog in the high-tech field. "TechCrunch was founded on June 11, 2005, as a weblog dedicated to obsessively profiling and reviewing new Internet products

and companies. In addition to covering new companies, we profile existing companies that are making an impact (commercial and/or cultural) on the new web space … TechCrunch has now grown into a network of technology focused sites offering a wide range of content and new media." Available online at http://techcrunch.com/about -techcrunch/#ixzz0mQNAPnHM.

37. Singularity University, a nonprofit institution founded in 2009, describes itself as "a unique interdisciplinary, international and intercultural experience which challenges students to use transformative, exponential technologies to address grand challenges. SU educates and inspires students to discover and design sustainable organizations to positively impact humanity." Available online at http://singularityu.org/faq/#VM1.

38. http://techcrunch.com/2010/03/14/craig-barrett-takes-on-vivek-wadhwa-in-the-tech -education-debate/.

39. As American Business for Legal immigration (ABLI). See http://www.monumentpolicy .com/.

40. Verdery's wife, Jenifer Eisen Verdery, is a lobbyist for Intel on these same issues, with the title of director of Public Policy. Previously she was the first executive director of *Compete America* when it was established in 1994 at the National Association of Manufacturers and, perhaps coincidentally, she also is the daughter of a then-vice president of that organization, Phyllis Eisen.

41. This description is quoted from the biography provided by Mr. Verdery to the *National Journal*, to which he is a contributor. Available online at http://lobbying.nationaljournal .com/contributors/c-stewart-verdery-jr.php.

42. Monument Policy Group, "Scott Corley, Respected Technology Advocate and Former Senate Advisor, to Join Monument Policy Group," Press Release October 7, 2009. Available online at http://www.mon0umentpolicy.com/scottcorleyannouncement.

43. Center for Responsible Politics, OpenSecrets.org, "Revolving Door." Available online at http://www.opensecrets.org/revolving/rev_summary.php?id=71574. Accessed February 11, 2013.

44. Narayanswamy, "Immigration: Give me your poor, your tired … your lobbyists?"

45. In 2001, Abramoff "bolted" with seven members of what he called "Team Abramoff" to become senior director of Governmental Affairs at another legal and lobbying practice in Washington, Greenberg Traurig. The addition of Abramoff's large lobbying billings quickly transformed Greenberg Traurig into one of the four or five largest lobbying firms in that city.

46. The "Gates" in the name of this prominent Seattle-based law firm was that of one of its co-founders, William Gates, Sr. The senior Gates is father of William Henry "Bill" Gates III, now non-executive Chairman of Microsoft. Abramoff describes William Gates, Sr. as the law firm's "senior partner." Although the account of a highly paid lobbyist and con-fessed felon may reasonably be viewed with doubt, it is worth noting that in his book Abramoff also claims that when he joined the law firm's Washington, DC subsidiary in December 1994, it represented primarily a single client, Microsoft Corporation. "The firm was so close to Microsoft that they sometimes operated like symbiotic organisms, nurtur-ing each other. The revolving door between the firm and Microsoft was constantly in motion. The head of legal affairs at the company had come from the firm. Attorneys at the firm had worked at Microsoft." Jack Abramoff, *Capitol Punishment: The Hard Truth about Washington Corruption from America's Most Notorious Lobbyist* (New York: WND Books, 2011), pp. 61–63. In 2007, Preston Gates & Ellis merged with Kirkpatrick & Lockhart Nicolson Graham to form K&L Gates LLP, now one of the largest global law firms. This firm's website lists Microsoft Corporation as one of its "representative clients." It also markets its Washington office as including "one of Washington's largest and most successful public policy and lobbying practices. Founded four decades ago, at a time

when few law firms had lobbying practices, the policy group has grown from a single lobbyist to become the largest of any fully integrated global law firm.… This group includes a bi-partisan team of nearly 50 attorneys and policy professionals with an in-depth knowledge of the legislative process, substantive experience and strong ties to key decision makers in Congress.… We have strong relationships with Democrats and Republicans in the House, the Senate and the Administration. We have over 500 years collective government experience. Among our attorneys and professionals are two former House members, and key former Republican and Democratic counsel and staff to the House and Senate leadership and committees." Available online at http://www.klgates.com/washington-dc-united-states-of-america/. Accessed August 14, 2013.

47. The Wiki website www.SourceWatch.org (downloaded September 9, 2010) reports as follows on William P. Jarrell: "Another Preston, Gates, and Ellis lobbying client was the Microsoft Corporation, per page 6 of the 1998 Mid-Year lobbying disclosure for Microsoft Corporation at *sopr.senate.gov*. Jarrell's specific lobbying activity included the successful expansion of the controversial H-1B visa program in 1998 via S.1723." Available online at http://sourcewatch.org/index.php?title=William_P._Jarrell; Sourcewatch.org is published by the Center for Media and Democracy, "an independent, non-profit, non-partisan media and consumer watchdog group." Available online at http://www.prwatch.org/cmd/index.html.

48. Gary Ater, "The Issues behind the Abramoff Scandal Started in 1957," *Common Sense*, February 24, 2007. Available online at http://commonsense-gater.blogspot.com/2007/02/issues-behind-abramhoff-scandal-started.html#!/2007/02/issues-behind-abramhoff-scandal-started.html.

49. Personal communication, Bruce A. Morrison, October 25, 2010. Congressman Morrison chaired the House Immigration Subcommittee from 1989 until 1991, when he ran unsuccessfully for Governor of Connecticut. From 1991 to 1997 he was a member of the U.S. Commission on Immigration Reform, and subsequently has continued to be involved in immigration policy issues as a Washington lobbyist. See "Tom Delay in the South Pacific," *Counterpunch*, June 15, 1999. Available online at http://www.counterpunch.org/1999/06/15/tom-delay-in-the-south-pacific/. The article also includes detailed quotes of statements made by DeLay during his visit to the Northern Marianas in 1999.

50. It is worth noting here that the Commonwealth of the Northern Marianas was one of the largest clients of "Team Abramoff." In July 1995 it retained Preston Gates Ellis & Rouvelas Meeds LLP to block congressional and executive branch efforts to harmonize its immigration and wage policies with those of the rest of the country, an effort that succeeded until 2007. Abramoff devotes a substantial part of his book to his lobbying on behalf of the Northern Marianas. See Abramoff, *Capitol Punishment*, 2011, pp. 66–82, 109–34. Jarrell's lobbying on behalf of the Northern Marianas also took place while he was working with Abramoff at Preston Gates. See "Tom Delay in the South Pacific," *Counterpunch*, June 15, 1999.

51. Susan Schmidt, James V. Grimaldi, and R. Jeffrey Smith, "Investigating Abramoff—Special Report," *Washington Post*. Available online at http://www.washingtonpost.com/wp-dyn/content/custom/2005/12/28/CU2005122801176.html. Accessed August 14, 2013.

52. The American Immigration Lawyers Association (AILA), an association of some 11,000 U.S. immigration lawyers, has been one of the most active lobbying organizations on immigration policy related to scientists and engineers. The much larger American Bar Association has its own committee on immigration law, but it is far less engaged in political activities and lobbying than is AILA. AILA has been politically astute in initiating or supporting other organizations with similar perspectives and placing its staff in key positions. As noted elsewhere, AILA was active in the creation of the corporate lobbying

organization now called *Compete America*, and a former AILA staff member Jenifer Eisen became the new organization's first executive director.

53. The Center for American Progress (CAP) is a well-funded and well-connected think tank and advocacy organization with a declared liberal/progressive agenda. It was created and still directed by John Podesta, former Chief of Staff to President Clinton. CAP reportedly has an annual budget of $25 million, but declines to make public a full list of its funders, for which it has been criticized by proponents of openness in government. According to a 2008 article in *Politico*, the "Center for American Progress, and its president, John Podesta, are uniquely integrated with the [Obama Administration] transition. Podesta, on leave from the Center for American Progress (CAP), heads the transition operation. The transition's operations director, general counsel, and co-director all shifted from similar jobs at CAP, and the transition is full of lower-level former CAP staffers or current board members." See Ben Smith and Chris Frates, "Where's transparency of Podesta Group?" *Politico*, December 8, 2008. Available online at http://www.politico.com/news/stories/1208/16318.html. CAP's current director of Immigration Policy, Marshall Fitz, previously was director of Advocacy at the American Immigration Lawyers Association.

54. In the interest of full disclosure, the author was an At-Large Member of the Board of Directors of the National Immigration Forum during its first two years.

55. See http://www.populationconnection.org/site/PageServer.

56. Only limited information available about the revenues of private lobbying firms, but Rick Swartz and Associates may be lucrative for its proprietor if one published estimate of its annual revenues—$500,000 to $1 million—is close to accurate: "Rick Swartz & Associates Inc in Washington, DC is a private company categorized under Lobbyists. Our records show it was established in 1990 and incorporated in District of Columbia. Current estimates show this company has an annual revenue of $500,000 to $1 million and employs a staff of approximately 1 to 4." Available online at http://www.manta.com/c/mm2zfsl/rick-swartz-associates- . Downloaded August 9, 2013.

57. John Heilemann, "Do you know the way to ban José?" *Wired Magazine* 4, no. 8 (August 1996), p. 7. Available online at www.wired.com/wired/archive/4.08/netizen_pr.html. Accessed August 14, 2013.

58. The Club for Growth lists its goals as follows: "Reduce income tax rates; Death tax repeal; Limited government through limited spending and budget reform, including a Balanced Budget Amendment to the United States Constitution; Social Security reform with personal retirement accounts for younger workers; Expanding trade freedom (free trade); End abusive lawsuits through medical malpractice and tort reform; Replacing the current tax code (flat tax, fair tax); School choice; Regulatory reform and deregulation." Available online at http://www.clubforgrowth.org/aboutus/. Accessed August 14, 2013.

59. Tom Barry, "Politics of class and corporations," CIP Americas, September 8, 2005. Available online at http://www.cipamericas.org/archives/1365. Accessed August 14, 2013.

60. Barry, "Politics of class and corporations"; Heilemann, "Do you know the way to ban José?" p. 7. Available online at www.wired.com/wired/archive/4.08/netizen_pr.html.

61. Full disclosure: the author was a member and vice chair of this Federal Commission.

62. Indeed, the former chair of the House Immigration Subcommittee, and original co-author of the 1990 provision creating the H-1B visa, believes the claims of influence by this Left-Right coalition have been much inflated. Bruce A. Morrison, personal communication, October 25, 2010.

63. Its awkward name is explained by its history: the association's original name was the National Association of Foreign Student Advisors (NAFSA), by which it became known in Washington. When it later decided to change its name to the "Association of International Educators," it retained the NAFSA acronym.

64. Nicholas Confessore, "Borderline Insanity: President Bush wants the INS to stop granting visas to terrorists. The biggest obstacle? His own administration," *Washington Monthly*, May 2002. Available online at http://www.washingtonmonthly.com/features/2001/0205 .confessore.html. Accessed August 14, 2013.

65. Confessore, "Borderline Insanity," no page numbers provided in online version.

66. NAFSA emphasizes that only one of the pilots originally entered the United States on a student visa. The other two pilots with student visas had entered the United States on other temporary visas and then applied for a change to student visa status. See Robert Farley, "9/11 Hijackers and Student Visas," FactCheck.org. Available online at http://news .yahoo.com/9-11-hijackers-student-visas-211835290.html.

67. All quotes from Confessore, "Borderline Insanity," n.p.

68. Government Accountability Office, *Student and Exchange Visitor Program: DHS Needs to Assess Risks and Strengthen Oversight Functions*, GAO-12-572, June 18, 2012. Available online at http://www.gao.gov/products/GAO-12-572.

69. Remarks by Kevin Casey, Associate Vice President, Office of Government, Community, and Public Affairs, Harvard University, NBER workshop on US Scientific Workforce, Cambridge, MA, January 24, 2002.

70. The National Education Association has 3.2 million members; the American Federation of Teachers has nearly 900,000 members.

71. Institute of Electrical and Electronics Engineers, "Return of Organization Exempt from Income Tax," Internal Revenue Service Form 990 for Fiscal Year 2009. Available online from IEEE website at http://www.ieee.org/about/financials.html.

72. Personal communication, Chris McManes, IEEE-USA Public Relations Manager, January 16, 2012.

73. Personal communication, Ron Hira, Rochester Institute of Technology, 2010.

74. See www.cssp.us.

75. See www.faseb.org.

Chapter 5: What Is the Market Really Like?

1. Carolyn M. Veneri, "Can Occupational Labor Shortages Be Identified Using Available Data?" *Monthly Labor Review* 122, no. 3 (March 1999), pp. 15–21, emphasis added. Citation on p. 15. Available online at http://www.bls.gov/opub/mlr/1999/03/art2full .pdf.

2. Burt S. Barnow, Jaclyn Schede, and John W. Trutko, *Occupational Labor Shortages: Concepts, Causes, Consequences, and Cures* (Kalamazoo, MI: W.E. Upjohn Institute for Employment Research, 2013), p. 2; emphasis added.

3. Such market adjustments may not take place in command economies such as those of the former Soviet Union, nor in those parts of the labor markets of mixed economies that are controlled or heavily influenced by government.

4. David M. Blank and George J. Stigler, *The Demand and Supply of Scientific Personnel* (New York: National Bureau of Economic Research, 1957), p. 23.

5. Kenneth J. Arrow and William M. Capron, "Dynamic Shortages and Price Rises: The Engineer-Scientist Case," *Quarterly Journal of Economics*, May 1959, p. 307.

6. For useful discussions of the economics concept of labor shortage, see Barnow et al., *Occupational Labor Shortage*, pp. 4–11. See also United Kingdom, Migration Advisory Committee, *Skilled, Shortage, Sensible: Review of Methodology*, March 2010, ch. 5, pp. 37–59. Available online at http://www.ukba.homeoffice.gov.uk/sitecontent/documents/aboutus/

workingwithus/mac/review-of-methodology/0310/mac-review-of-methodology?view =Binary.

7. Arrow and Capron, "Dynamic Shortages and Price Rises," p. 292.

8. National Research Council, Office of Scientific and Engineering Personnel, *Forecasting Demand and Supply of Doctoral Scientists and Engineers: Report of a Workshop on Methodology* (Washington, DC: National Academy Press, 2000), pp. 15–21.

9. National Research Council, *Forecasting Demand and Supply*, pp. 19–22.

10. National Science Board, *Science and Engineering Indicators 2012* (Arlington, VA: National Science Foundation, 2012 [NSB 12-01]), fig. 3.3, pp. 3–11. Available online at http://www .nsf.gov/statistics/seind12/pdf/c03.pdf.

11. National Science Foundation, "About SESTAT: Who is included in the definition of 'scientist' and 'engineer' in SESTAT?" Available online at http://www.nsf.gov/statistics/ sestat/sestatfaq.cfm#question003.

12. National Science Board, *Science and Engineering Indicators 2012* (Arlington, VA: National Science Foundation, 2012 [NSB 12-01]), table 3.3, pp. 3–10. Available online at http:// www.nsf.gov/statistics/seind12/pdf/c03.pdf.

13. National Science Board, *Science and Engineering Indicators 2012*. Arlington VA: National Science Foundation, 2012 [NSB 12-01]), table 3.3, pp. 3–10. Available online at http:// www.nsf.gov/statistics/seind12/pdf/c03.pdf.

14. Public discussion is further confused by often loose usage of other terms such as "STEM occupations," which unfortunately has no agreed definition. (STEM is an acronym that was created by the NSF to stand for Science, Technology, Engineering, Mathematics.) The National Science Board notes in passing that it considers this term "STEM occupations" to represent an expansion of the established list of science and engineering occupations to include several other large occupations: science and engineering technicians (a large category that commonly do not require bachelor's level degrees); science and engineering managers; and computer programmers (another large category that also includes many without degrees). See National Science Board, *Science and Engineering Indicators 2012* (Arlington, VA: National Science Foundation, 2012 [NSB 12-01]), pp. 3–7.

15. For a more complete discussion, see Anthony P. Carnevale, Nicole Smith, and Michelle Melton, *STEM: Science, Technology, Engineering, Mathematics* (Washington, DC: George-town University Center on Education and the Workforce, 2011), pp. 40–62. Available online at http://cew.georgetown.edu/stem/. Accessed August 14, 2013.

16. http://www.nsf.gov/statistics/srvyrecentgrads/surveys/srvyrecentgrads_2006.pdf.

17. Personal communication, Nirmala Kannankutty, Senior Advisor, National Center for Science and Engineering Statistics, National Science Foundation, September 2009 and December, 2011.

18. In 2009 the NSF began a project to clarify what respondents to such questions actually mean about technical expertise acquired from such a degree. Findings from this project are still only preliminary, but further analyses are planned. Personal communication, Nirmala Kannankutty, Senior Advisor, National Center for Science and Engineering Statistics, National Science Foundation, January 20, 2012.

19. Daniel E. Hecker, "High-technology Employment: A NAICS-based Update," *Monthly Labor Review* (July 2005), pp. 57–72, table 1. The lower levels are "Level II," with about 3 to 5 times the average, and "Level III," with about 2 to 3 times the average intensity of "technology-oriented workers." These industries are more focused on extraction (e.g., oil and gas extraction, forestry) and manufacturing (e.g., industrial machinery and chemical manufacturing), and are not normally included in discussions of industries that are "high-tech."

20. If all three levels are combined, the total was about 14.4 million, or 11 percent. Hecker, "High-technology Employment," table 1.

21. Philip Martin and Martin Ruhs, "Labor Shortages and U.S. Immigration Reform: Promises and Perils of an Independent Commission," *International Migration Review* 45, no. 1 (Spring 2011), pp. 174–87, citation is on p. 180.
22. Veneri, "Can Occupational Labor Shortages Be Identified," p. 15.
23. Veneri, "Can Occupational Labor Shortages Be Identified," p. 18.
24. Veneri, "Can Occupational Labor Shortages Be Identified," pp. 18–19.
25. Formerly long-time executive director of the Engineering Workforce Commission.
26. All can be accessed online at www.cpst.org. Full disclosure: the STEM Workforce Data Project was supported by a grant from the Alfred P. Sloan Foundation, of which the author was a vice president.
27. Richard A. Ellis, "Is U.S. Science and Technology Adrift?" STEM Workforce Data Project, Report No. 8 (Washington, DC: Commission on Professionals in Science and Technology, 2007), pp. 2–3. Available online at www.cpst.org.
28. William P. Butz, Gabrielle A. Bloom, Mihal E. Gross, Terrence K. Kelly, Aaron Kofner, and Helga E. Rippen, "Is There a Shortage of Scientists and Engineers? How Would We Know?" *RAND Issue Paper*, 2003, pp. 1–8.
29. William P. Butz, "Rapporteur's Summary," in *The U.S. Scientific and Technical Workforce: Improving Data for Policymaking*, ed. Terrence K. Kelly, William P. Butz, Stephen Carroll, David M. Adamson, and Gabrielle Bloom, Conference Proceedings prepared for the Office of Science and Technology Policy and the Alfred P. Sloan Foundation, RAND: June 2004, CF-194-OSTP/SF, pp. 102–3. Available online at www.rand.org/pubs/conf _proceedings/CF194.html.
30. Titus Galama and James Hosek, *U.S. Competitiveness in Science and Technology*, prepared for the Office of the Secretary of Defense, RAND National Defense Research Institute, 2008, p. 97. Available online at http://www.rand.org/pubs/monographs/MG674.html.
31. Galama and Hosek, *U.S. Competitiveness*, p. 100. Available online at http://www.rand.org/ pubs/monographs/MG674.html.
32. Galama and Hosek, *U.S. Competitiveness*, pp. 119–20. Available online at http://www.rand .org/pubs/monographs/MG674.html.
33. The NBER is a private, nonprofit, nonpartisan research organization founded in 1920 and "dedicated to promoting a greater understanding of how the economy works.... The NBER is the nation's leading nonprofit economic research organization." Its associates have included 22 Nobel Prize winners and 13 past chairs of the President's Council of Economic Advisers. Current NBER Associates include more than 1,100 selected professors of economics and business on the faculties of colleges and universities in North America. National Bureau of Economic Research, "History of the NBER." Available online at http://nber.org/info.html.
34. Richard B. Freeman, "What Does Global Expansion of Higher Education Mean for the United States? in *American Universities in a Global Market*, ed. Charles T. Clotfelter (Chicago: University of Chicago Press, 2010), pp. 373–404.
35. Freeman, "What Does the Global Expansion," p. 373.
36. Freeman, "What Does the Global Expansion," p. 377, table 11.3.
37. Freeman, "What Does the Global Expansion," p. 378, table 11.4.
38. Freeman, "What Does the Global Expansion," pp. 383–84.
39. Richard B. Freeman, "Does Globalization of the Scientific/Engineering Workforce Threaten U.S. Economic Leadership?" NBER Working Paper No. 11457, June 2005.
40. Freeman, "What Does the Global Expansion," pp. 399–400.
41. George Borjas, "Immigration in High-Skill Labor Markets: The Impact of Foreign Students on the Earnings of Doctorates," NBER Working Paper No. 12085, March 2006.
42. Paula Stephan, *How Economics Shapes Science* (Cambridge, MA and London, UK: Harvard University Press, 2012), p. 164.

43. Stephan, *How Economics Shapes Science*, p. 153.

44. Stephan, *How Economics Shapes Science*, pp. 151–52.

45. Stephan, *How Economics Shapes Science*, p. 153.

46. Giovanni Peri, "Rationalizing U.S. Immigration Policy: Reforms for Simplicity, Fairness, and Economic Growth," The Hamilton Project, Discussion Paper 2012-01 (Washington, DC: Brookings Institution, May 2012). Available online at http://www.hamiltonproject .org/files/downloads_and_links/05_peri_discussion.pdf. See also Pia Orrenius and Madeline Zavodny, "Foreign Stimulus," op-ed published in the *New York Times*, September 13, 2010. Available online at http://www.nytimes.com/2010/09/14/opinion/14orrenius .html?_r=0.

47. The term "diversion" may carry confusing connotations. In normal usage its root is a transitive verb that suggests deliberate action or intent "to divert," whereas it is clear that the authors have in mind a more market-driven process in which those graduating with STEM degrees find more attractive career paths in non-STEM occupations.

48. Anthony P. Carnevale, Nicole Smith, and Michelle Melton, *STEM: Science, Technology, Engineering, Mathematics* (Washington, DC: Center for Education and the Workforce, Georgetown University, 2011), p. 10. Available online at http://www9.georgetown.edu/ grad/gppi/hpi/cew/pdfs/stem-complete.pdf.

49. Joe Light and Rachel Emma Silverman, "Generation Jobless: Students Pick Easier Majors Despite Less Pay," *Wall Street Journal*, November 9, 2011. Available online at http://online .wsj.com/article/SB10001424052970203733504577026212798573518.html?KEYWORDS =anthony+carnevale+you%27d+have+to+be+crazy.

50. Carnevale et al., "STEM," p. 10

51. B. Lindsay Lowell and Hal Salzman, "Into the Eye of the Storm: Assessing the Evidence on Science and Engineering Education, Quality, and Workforce Demand," unpublished manuscript dated October 2007, p. 43. http://www.urban.org/url.cfm?ID=411562. The research for this report was supported by a grant from the Alfred P. Sloan Foundation.

52. Xie and Killewald, *Is American Science in Decline?*

53. Xie and Killewald, *Is American Science in Decline?* p. 61.

54. Xie and Killewald, *Is American Science in Decline?* pp. 42–43.

55. Xie and Killewald, *Is American Science in Decline?* pp. 116–19.

56. Xie and Killewald, *Is American Science in Decline?* p. 35.

57. "A Comprehensive Statistical Study of Education Trends in Three Disciplines Closely Linked to Energy-Producing Industries—Mining, Nuclear and Petroleum Engineering," *Engineering Trends*. Available online at http://engtrends.com/IEE/1107C.php.

58. *Gathering Storm*, p. 78.

59. Calculated from *Science and Engineering Indicators 2012*, appendix tables 2-18, 2-26, and 2-28.

60. See, for example, Jacqueline Ruttimann, "Positions Available: No Ph.D. Required?" *ScienceCareers*, August 15, 2008. Available online at August 15, 2008, http://sciencecareers .sciencemag.org/get-file.xqy?uri=/aaas/files/original/application/161c6be7e302da2c254d a62c8ceedc51.pdf.

61. National Academy of Engineering and National Research Council, Committee on Science, Technology, Engineering, and Mathematics Workforce Needs for the U.S. Department of Defense and the U.S. Defense Industrial Base, *Assuring the U.S. Department of Defense a Strong Science, Technology, Engineering, and Mathematics (STEM) Workforce* (Washington, DC: National Academies Press, 2012), ch. 3, especially p. 54. Available online at http://www.nap.edu/catalog.php?record_id=13467. The author served as a member of this committee.

62. "The PhD Factory: The World Is Producing More PhDs Than Ever Before. Is It Time To Stop?" *Nature*, 472 (April 21, 2011), pp. 276–77.

63. National Research Council, Board on Higher Education and Workforce, Policy and Global Affairs, Committee on an Assessment of Research Doctorate Programs, Jeremiah P. Ostriker, Charlotte V. Kuh, and James A. Voytuk, eds., *A Data-Based Assessment of Research-Doctorate Programs in the United States with CD*, 2011 (Washington, DC: National Academies Press, 2011). Available online at http://www.nap.edu/rdp/.

64. "Fix the PhD," Editorial, *Nature* 472 (April 21, 2011), pp. 259–60.

65. Xie and Killewald, *Is American Science in Decline?* table 4.2, p. 59.

66. Biomedical Research Workforce Working Group Report, Advisory Committee to the Director, National Institutes of Health, June 14, 2012, p. 19, fig. 3. Available online at http://acd.od.nih.gov/bwf.htm.

67. See Richard Freeman, Eric Weinstein, Elizabeth Marincola, Janet Rosenbaum, and Frank Solomon, "Careers and Rewards in Bio Sciences: The Disconnect between Scientific Progress and Career Progression." (Bethesda, MD: American Society for Cell Biology, 2001), pp. 14–15. Available online at http://www.ascb.org/newsfiles/careers_rewards.pdf. See also Stephan, *How Economics Shapes Science*, p. 157, Table 7.1.

68. Stephan, *How Economics Shapes Science*, p. 161.

69. Richard Monastersky, "The Real Science Crisis: Bleak Prospects for Young Researchers," *Chronicle of Higher Education*, September 21, 2007. Available online at http://chronicle .com/article/The-Real-Science-Crisis-Bleak/29178/.

70. "The PhD Factory," pp. 276–79.

71. Editorial, *Nature* 472 (April 21, 2011), p. 260.

72. Biomedical Research Workforce Working Group Report, Advisory Committee to the Director, National Institutes of Health, June 14, 2012. Available online at http://acd.od .nih.gov/bwf.htm. The author served as a member of the Modeling Subcommittee of the Working Group.

73. Presentation, Biomedical Research Working Group to meeting of the Advisory Committee to the Director, National Institutes of Health, June 14, 2012. Available online at http:// acd.od.nih.gov/reports/june2012-BMW-working-group-508.PDF.

74. The oral presentation was webcast by the NIH and is easily accessible from the NIH video archive at http://videocast.nih.gov/launch.asp?17313. The cited comments by Drs. Tilghman and Collins may be found at approximately minutes 167 and 245 respectively.

75. B. Lindsay Lowell, "Temporary Visa for Highly Skilled Jobs: A Brief Overview of the Specialty Worker H-1B Visa," Presentation at a Policy Briefing on Temporary Worker Programs, Institute for the Study of International Migration, Georgetown University, March 28, 2008. See especially the figure entitled "Specialty H-1B Entrants and Population," and fig. 2, "The H-1B Workforce in Information Technology (IT) and US IT Imports." Available online at http://www12.georgetown.edu/sfs/isim/Event%20 Summaries&Speeches/Sloan/Lowell,%20TWPBriefing%20H1Bs%203-28-08.pdf.

76. See Commission on Professionals in Science and Technology, "Postdocs and Career Prospects: A Status Report" (Washington, DC: Commission on Professionals in Science and Technology, June 1997). http://www.cpst.org/web/site/pages/pubs/pubpage.cfm.

77. Letter dated January 29, 2007 to Alyson Reed, executive director of National Postdoctoral Association from Norka Ruiz Bravo, PhD, deputy director for Extramural Research, National Institutes of Health, and Kathie L. Olsen, PhD, deputy director, National Science Foundation. Copy available on NIH website at http://grants1.nih.gov/training/ Reed_Letter.pdf.

78. Commission on Professionals in Science and Technology, "Postdocs and Career Prospects: A Status Report" (Washington, DC: CPST, 1997), 26pp.http://books.google .com/books/about/Postdocs_and_career_prospects.html?id=wpAFHQAACAA

79. For a more detailed discussion, see National Science Board, *Science and Engineering Indicators, 2008*, pp. 3–35 to 3–37. Available online at http://www.nsf.gov/statistics/ seind08/pdf/c03.pdf.

80. National Science Board, *Science and Engineering Indicators 2012* (Arlington VA: National Science Foundation [NSB 12-01]), pp. 3–33. Available online at http://www.nsf.gov/statistics/seind12/pdf/c03.pdf.

81. Xie and Killewald, *Is American Science in Decline?* table 4.2, p. 59. The authors compared earnings for lawyers with those for PhD-level scientists and engineers, presumably because the standard professional degree for lawyers in the United States is now called the *Juris Doctor*, or JD. It is worth mentioning, however, that the standard three year professional law degree was called a "Bachelor of Laws" or LL.B. until 50 years ago; that the JD degree is still a three-year post-baccalaureate degree, far shorter than the typical PhD; and that the two-year post-baccalaureate professional degree in business is generally called a Master's of Business Administration, or MBA. It may be, then, that the more appropriate comparison for lawyers with scientists and engineers would be at the master's rather than the PhD level, although this would only increase the earnings advantage of lawyers.

82. Stephan, *How Economics Shapes Science*, p. 155, fig. 7.2.

83. Data derived from Stephan, *How Economics Shapes Science*, p. 155, fig. 7.2.

84. Richard A. Ellis, "Is U.S. science and technology adrift?" *STEM Workforce Data Project*, *Report No. 8* (Washington, DC: Commission on Professionals in Science and Technology, 2007), pp. 2–3. Available online at www.cpst.org.

85. Ellis, "Is U.S. Science and Technology Adrift?" p. 3.

86. Paul Kedrosky and Dane Stangler, "Financialization and Its Entrepreneurial Consequences," Kansas City, MO: Kauffman Foundation Research Series: Firm Formation and Economic Growth, March 2011, pp. 6–7.

Chapter 6: The Distinctive U.S. Academic Production Process

1. Testimony before the Committee on Financial Services, U.S. House of Representatives, February 15, 2006.

2. China has been rapidly expanding its higher education system and is now said to have more than 2,000 universities and colleges. Japan has fewer than 200 public universities, but there are also more than 500 private four-year colleges and universities, which makes the Japanese higher education system look somewhat more like that of the United States than those in Europe.

3. A few institutions of higher education are supported directly by the federal government, mostly related to the military. These include the United States Military Academy at West Point, the U.S. Naval Academy at Annapolis, the U.S. Air Force Academy at Colorado Springs, the National Defense University, and the Naval Postgraduate School. Most of these are primarily undergraduate institutions, and their students are primarily active-duty military personnel. Another partial exception is Howard University in Washington, DC, an historically black research university (HBCU) that is federally chartered and is the beneficiary of an unusual line-item congressional appropriation in the budget of the U.S. Department of Education, which in FY2008 provided $234 million.

4. For a full list of the 62-member institutions of the AAU, which includes two Canadian universities, see http://aau.edu/about/article.aspx?id=5476.

5. http://aau.edu/about/article.aspx?id=5474.

6. There are also some federal agencies that support research in the humanities, arts, and social sciences, (e.g., the National Endowment for the Arts, the National Endowment for the Humanities, the National Science Foundation, etc.). However, by far the bulk of this federal research funding is directed toward natural science and engineering.

7. Times Higher Education (London), "The World University Rankings 2011–2012," available online at http://www.timeshighereducation.co.uk/world-university-rankings/. The rankings are based on assessments along the following dimensions:

Teaching — the learning environment (worth 30 percent of the overall ranking score)

Research — volume, income, and reputation (worth 30 percent)

Citations — research influence (worth 30 percent)

Industry income — innovation (worth 2.5 percent)

International outlook — staff, students, and research (worth 7.5 percent).

8. Center for World-Class Universities, Shanghai Jiao Tong University, "The Academic Ranking of World Universities" (Shanghai, China: ShanghaiRanking Consultancy, 2011). Available online at http://www.shanghairanking.com/index.html.

9. Vannevar Bush, *Science: The Endless Frontier*, A Report to the President by the Director of the Office of Scientific Research and Development, July 1945 (Washington, DC: U.S. Government Printing Office, 1945). Full text available online at http://www.nsf.gov/about/history/vbush1945.htm.

10. Here is I. B. Cohen's first-hand account:

Let me conclude this eyewitness report with a final observation. *Science: The Endless Frontier* was produced in response to a letter of request addressed by Franklin D. Roosevelt to Vannevar Bush. Whose idea was it? Who wrote the letter?

I conducted an oral-history interview with Vannevar Bush a few years before his death. I asked him straight out, ... whose idea was it to commission such a report? Who had written the letter? He looked me in the eye, and without a moment's hesitation said the idea was his. He turned his head a little bit to one side as was his habit, smiled, stated unequivocally, "I wrote the letter."

There was no occasion for further discussion.

From Consortium for Science, Policy and Outcomes, Arizona State University, "Science The Endless Frontier 1945–1995: Learning from the Past, Designing for the Future, Seminar transcript," Part I—December 9, 1994, p. 23. http://www.cspo.org/products/conferences/bush/partone/complete94.pdf.

11. In the 1930s, Bush also had founded a company that later evolved into Raytheon Corporation, still one of the prime defense contractors.

12. See Irvin Stewart, *Organizing Scientific Research for War: The Administrative History of the Office of Scientific Research and Development* (Boston: Little Brown, 1948), esp. ch. 3. Available online at http://archive.org/details/organizingscient00stew.

13. Bush, *Science: The Endless Frontier*, ch. 3.

14. See for example Shirley Ann Jackson, "Science the Endless Frontier: The Continuing Relevance of Vannevar Bush," The Green Center for Physics Dedication, Massachusetts Institute of Technology, Friday, October 5, 2007. http://www.rensselaer.edu/president/speeches/ps100507-mitgreencenter.html.

15. Bush, *Science: The Endless Frontier*, chapters 1 and 3, n.p. Available online at http://www.nsf.gov/od/lpa/nsf50/vbush1945.htm.

16. These actions were taken in part due to findings of NSF-supported research by Freeman, Chang, and Chiang that concluded that:
raising the value of awards increases the number of applicants and quality of awardees while giving more awards increases the number of awardees, by

definition, with only a modest reduction in measured academic skills ... [and]
more and better-paying stipends could raise the number of native-born/residents
choosing S&E fields broadly.

From Richard Freeman, Tanwin Chang, Hanley Chiang, "Supporting 'The Best and
Brightest' in Science and Engineering: NSF Graduate Research Fellowships" (Cambridge,
MA: National Bureau of Economic Research and Harvard University, revised March
2006), p. 27. Available online at http://www.nber.org/~sewp/freeman_nsfstip_proceedings
.pdf.

17. National Research Council, Committee to Study the National Needs for Biomedical,
Behavioral, and Clinical Research Personnel, *Research Training in the Biomedical,
Behavioral, and Clinical Research Sciences* (Washington, DC: National Academies Press,
2011), p. 14. Available online at http://www.nap.edu/catalog.php?record_id=12983.

18. National Research Council, *Research Training*, p. 13.

19. National Science Foundation, "Number of People Involved in NSF Activities," FY 2011
NSF Budget Request to Congress, Summary Tables, table 12. Available online at http://
www.nsf.gov/about/budget/fy2011/pdf/02-Summary_Tables_&_Charts_fy2011.pdf.

20. National Research Council, *Research Training*, p. 13.

21. See, for example, National Research Council, Committee on Dimensions, Causes, and
Implications of Recent Trends in the Careers of Life Scientists, *Trends in the Early Careers
of Life Scientists* (Washington, DC: National Academies Press, 1998).

22. National Research Council, *Research Training*, p. 17.

23. Stephan, *How Economics Shapes Science*, pp. 161–63.

24. Paula Stephan, "Tracking the Placement of Students as a Measure of Technology Transfer,"
in *Advances in the Study of Entrepreneurship, Innovation, and Economic Growth*, ed. Gary
Libecap (London: Elsevier, 2009), pp. 113–40. See also Stephan, *How Economics Shapes
Science*, p. 162.

25. Quotations are from U.S. Department of State, "J-1 Exchange Visitor Program." Available
online at http://j1visa.state.gov/programs.

26. National Science Board, *Science and Engineering Indicators 2012*, pp. 2–9, table 2.7.

27. National Science Board, *Science and Engineering Indicators 2012*, pp. 2–25. See also
appendix table 2-24.

28. National Science Board, *Science and Engineering Indicators 2012*, pp. 2–26. See also
National Science Board, *Science and Engineering Indicators 2010*, http://www.nsf.gov/
statistics/seind10/c2/c2s3.htm.

29. Karen Kaplan, "Postdoc or Not?" *Nature* 483 (March 22, 2012), pp. 499–500.

30. Consider the following 1992 interview comment by Dr. James Watson, the Nobel
Laureate who co-discovered the structure of DNA at the age of 24:

Interviewer: Going back to your position at 24, making the discovery that would
win the Nobel Prize, do you think today that biology is a field for young scientists?

Dr. James Watson: I think you're unlikely to make an impact unless you get into a
really important lab at a young age, because you're unlikely to know what problem
to work on ... People used to be kings when they were nineteen, generals. Now
you're supposed to wait until you're virtually senile. In fact, we sometimes choose
senility because it doesn't threaten anyone.

From Interview with Dr. James Watson, "Winding Your Way through DNA" symposium,
University of California San Francisco, 1992. Available online at http://www.access
excellence.org/RC/CC/watson.php.

31. A related example of incoherence is that the numbers of H-1B visas provided to both
corporate and nonprofit employers are far larger than the numbers of permanent visas
for would-be immigrants with high levels of education and skills, leading to large
backlogs of former H-1B visa-holders awaiting permanent visas.

32. Richard B. Freeman, "Does Globalization of the Scientific/Engineering Workforce Threaten U.S. Economic Leadership?" (Cambridge, MA: National Bureau of Economic Research, June 2005), Working Paper 11457, Abstract. Available online at www.nber.org/papers/w11457.
33. Xie and Killewald, *Is American Science in Decline?* pp. 100, 123.
34. Bush, *Science: The Endless Frontier*, ch. 4.

Chapter 7: International Comparisons: Glass Half Full, Glass Half Empty

1. *Affurisms from Josh Billings: His Sayings*, 1865. "Josh Billings" was the pen name of the popular American humorist Henry Wheeler Shaw (1818–1885).
2. For a useful compendium of such arguments, see the website of the Alliance for Science and Technology Research in America (ASTRA), which describes itself as "a collaboration of over 130 companies, academic institutions, professional societies, trade associations, and foundations. ASTRA's mission is to ensure that there is an adequate, and growing, investment by the federal government in basic research in the physical sciences, the mathematical and computational sciences, and engineering. To accomplish this goal, ASTRA has developed facts-based advocacy and research programs and publications to educate and inform the general public about the relationship between R&D investments and resulting benefits to the U.S." Available online at http://www.aboutastra.org/about/index.asp. See also Council on Foreign Relations, "Expert Roundup: U.S. Innovation and Economic Recovery," June 6, 2011. Available online at http://www.cfr.org/economics/us-innovation-economic-recovery/p25198.
3. Members of the 2005 "Rising Above the Gathering Storm" Committee, *Rising Above the Gathering Storm, Revisited: Rapidly Approaching Category 5* (Washington, DC: National Academies Press, 2005), p. 5. Available online at http://www.nap.edu/catalog.php?record_id=12999.
4. Members of the 2005 "Rising Above the Gathering Storm" Committee, *Rising Above the Gathering Storm, Revisited*, pp. 6–15. http://www.nap.edu/catalog.php?record_id=12999.
5. National Science Board, *Science and Engineering Indicators 2012*, p. 0-3.
6. Mark Boroush, "U.S. R&D Spending Suffered a Rare Decline in 2009 but Outpaced the Overall Economy," National Science Foundation InfoBrief, NSF 12-310, March 2012, p. 7.
7. Calculated from Boroush, "U.S. R&D Spending," table 4, p. 8. All national data are expressed in U.S. dollars based upon purchasing power parities (see note 4 in Boroush text).
8. Baroush, "U.S. R&D Spending," table 4, p. 8.
9. National Science Board, *Science and Engineering Indicators 2012*, p. 0-4.
10. Mark Baroush, "New NSF Estimates Indicate that U.S. R&D Spending Continued to Grow in 2008," National Science Foundation InfoBrief, NSF 10-312, January 2010, p. 3.
11. Baroush, "New NSF Estimates," p. 4.
12. See, for example, Ralph Gomory, "The Innovation Illusion," *Manufacturing and Technology News* 17, no. 9 (May 31, 2010). Available online at http://www.manufacturingnews.com/news/10/0531/gomory.html.
13. National Science Board, *Science and Engineering Indicators 2012*, p. 0-3.
14. National Science Board, *Science and Engineering Indicators 2012*, p. 0-5.
15. *National Science Board, Science and Engineering Indicators 2010*, footnote 7, p. 0-20, and fig. 0-7, p. 0-7.

16. Richard B. Freeman, "Does Globalization of the Scientific/Engineering Workforce Threaten U.S. Economic Leadership?" Working Paper 11457, National Bureau of Economic Research, 2005, p. 3. Available online at http://www.nber.org/papers/w11457

17. National Science Board, *Science and Engineering Indicators 2010*, fig. 0-7, p. 0-7. Organization for Economic Cooperation and Development (OECD), *Education at a Glance 2011: OECD Indicators* (Paris: OECD Publishing, 2011). Available online at http://dx.doi.org/10.1787/eag-2011-en.

18. The G-20 group includes the world's largest economies—19 countries plus the European Union, which now includes 28 countries of which four also are members of the G-20 group. The member countries of the G-20 are: Argentina, Australia, Brazil, Canada, China, France, Germany, India, Indonesia, Italy, Japan, Mexico, Russian Federation, Saudi Arabia, South Africa, Republic of Korea, Turkey, United Kingdom, and the United States. The European Union is represented by the rotating Council presidency and the European Central Bank. Data for two of the large G-20 countries, India and the Russian Federation, are unavailable and hence not included in these calculations.

19. Calculated from Organization for Economic Cooperation and Development (OECD), *Education at a Glance 2011: OECD Indicators* (Paris: OECD Publishing, 2011, chart A1.4, p. 35 and table A1.3a, p. 40. Available online at http://dx.doi.org/10.1787/eag-2011-en. See also National Science Board, *Science and Engineering Indicators 2010*, fig. 0-7, p. 0-7. Due to the rapid expansion of higher education in China since 2000, its true percentages for 2009 were likely larger and those of other countries somewhat smaller.

20. Joan Burelli and Alan Rapoport, "Reasons for International Changes in the Ratio of Natural Science and Engineering Degrees to the College-Age Population," National Science Foundation InfoBrief 09-308, January 2009, p. 3, table 2. This table includes similar data for many other countries.

21. Burelli and Rapoport, "Reasons for International Changes," p. 3, table 2.

22. Burelli and Rapoport, "Reasons for International Changes," pp. 1, 6.

23. Percentages have been calculated from National Science Board, *Science and Engineering Indicators 2012*, appendix table 2-32.

24. National Science Board, *Science and Engineering Indicators 2012*, appendix table 2-32.

25. See, for example, Gary Gereffi, Vivek Wadhwa, Ben Rissing, and Ryan Ong, "Getting the Numbers Right: International Engineering Education in the United States, China, and India," *Journal of Engineering Education* 97, no. 1 (2008), pp. 13–25. Available online at http://papers.ssrn.com/sol3/papers.cfm?abstract_id=1081923. Also see Freeman, "Does Globalization," available online at http://www.nber.org/papers/w11457.

26. Calculated from data in John H. Pryor, Kevin Eagan, Laura Palucki Blake, Sylvia Hurtado, Jennifer Berdan, and Matthew H. Case, *The American Freshman: National Norms Fall 2012* expanded edition (Los Angeles: Higher Education Research Institute, Graduate School of Education & Information Studies, University of California, Los Angeles, 2012), appendix A, pp. 26–27. Available online at http://heri.ucla.edu/monographs/TheAmericanFreshman2012-Expanded.pdf.

27. National Science Board, *Science and Engineering Indicators 2012*, ch. 2, pp. 2-22 to 2-23.

28. Hal Salzman, "Reconceptualizing Science and Engineering Pathways: An Empirical Analysis of College Student Flows, or Is Discipline Destiny?" Rutgers University, J. J. Heldrich Center for Workforce Development, Working Paper 2013, 18pp.

29. To minimize confusion we have separated out the complex NSF category "Social/behavioral sciences," which accounts for a further 8 percent of entering students but includes some large majors such as history that arguably are not in the "science" category; psychology as a mixture of natural and social science sub-disciplines; and sociology, economics, and political science among others.

30. Calculated from National Science Board, *Science and Engineering Indicators 2012*, ch. 2, table 2.9.
31. Unlike the NSB analysis, Salzman included pre-professional degrees such as pre-med, nursing, and physical therapy as part of S&E disciplines, on grounds that their curricula include a full complement of science courses comparable to biology degrees. Salzman, "Reconceptualizing," pp. 6–7.
32. Salzman, "Reconceptualizing," p. 9.
33. Salzman, "Reconceptualizing," pp. 13–14.
34. Calculated from National Science Board, *Science and Engineering Indicators 2012*, appendix table 2-32.
35. Calculated from National Science Board, *Science and Engineering Indicators 2012*, appendix table 2-32.
36. Times Higher Education Supplement, "World University Rankings, 2011–2012." Available online at http://www.timeshighereducation.co.uk/world-university-rankings/2012-13/world-ranking.
37. Center for World Class Universities of Shanghai Jiao Tong University, "Academic Rankings of World Universities," 2013. Available online at http://www.shanghairanking.com/index.html#.
38. "The PhD Factory," *Nature* 472 (April 21, 2011), pp. 276–29. Quotes from p. 276.

Chapter 8: Making Things Work Better

1. Joseph E. Stiglitz, "Knowledge as a Global Public Good," in *Global Public Goods*, Inge Kaul, Isabelle Grunberg, and Marc A. Stern, eds. (New York and Oxford: Oxford University Press, 1999), pp. 308–25. Available online at http://web.undp.org/global publicgoods/TheBook/globalpublicgoods.pdf.
2. These included the Atomic Energy Commission (AEC), the Energy Research and Development Administration, and the regulatory agency known as the Federal Power Commission. The AEC was a civilian agency established after World War II to assume leadership of the nation's nuclear weapons programs that evolved out of the military's Manhattan Project during World War II, along with the development and regulation of civilian nuclear power.
3. The formative role of the Department of Energy in what became the successful Human Genome Project is sometimes overlooked, in part because leadership and primary funding of that project shifted to the National Institutes of Health. See U.S. Department of Energy, "Major Events in the U.S. Human Genome Project and Related Projects." Available online at http://www.ornl.gov/sci/techresources/Human_Genome/project/timeline.shtml.
4. Bush, *Science: The Endless Frontier*, ch. 6.
5. NIH's research budget is more than four times that of the National Science Foundation. Among U.S. government agencies, the NIH budget is exceeded only by the R&D budget of the Department of Defense, but the latter is heavily concentrated upon the "D" or development side of R&D. For an informative overview, see Congressional Research Service, *Federal Research and Development Funding: FY2011*, March 10, 2010. Available online at http://assets.opencrs.com/rpts/R41098_20100310.pdf.
6. David Korn et al., "The NIH Budget in the 'Postdoubling' Era," *Science* 296 (May 24, 2002), p. 1402.

7. Henry R. Bourne, "The Biomedical Workforce Report (TRR-I)," July 26, 2012, p. 2. Available online at http://biomedwatch.wordpress.com/2012/07/26/the-biomedical-workforce-report-trr-i/.

8. See, for example, Francis S. Collins, "Opportunities for Research and NIH," *Science* (January 1, 2010),pp. 36–37; and Brian C. Martinson, The Academic Birth Rate," *EMBO Reports* 12, no. 8 (2011), pp. 758–62.

9. There is much evidence on this subject. For well-informed assessments of the most recent evidence, see Stephan, *How Economics Shapes Science*, pp. 152–82; and Xie and Killewald, *Is American Science in Decline?* pp. 57–63. See also Donald Kennedy, Jim Austin, Kirstie Urquhart, and Crispin Taylor, "Supply Without Demand," *Science* (February 20, 2004), p. 1105.

10. See Biomedical Research Workforce Working Group Report, Advisory Committee to the Director, National Institutes of Health, June 14, 2012. Available online at http://acd.od.nih.gov/Biomedical_research_wgreport.pdf.

11. Michael S. Teitelbaum, "Structural Disequilibria in Biomedical Research," *Science* 321 (August 1, 2008), pp. 644–45.

12. The section on perverse incentives is based on a memorandum prepared by the author for the Biomedical Workforce working group of the Advisory Committee to the Director, National Institutes of Health, and published as Appendix D of that working group's report. See "Incentives Memorandum," in Biomedical Research Workforce Working Group Report, Advisory Committee to the Director, National Institutes of Health, June 14, 2012, appendix D, pp. 72–80. Available online at http://acd.od.nih.gov/Biomedical_research_wgreport.pdf.

13. There is however a cap on the overall amount of salary that can be paid to an individual faculty member by NIH. Beginning in late 2012, the *Consolidated Appropriations Action, 2012* (Public Law 112-74) signed into law on December 23, 2011 lowered the salary limitation on NIH Grants from Executive Level I ($199,700) to Executive Level II ($179,700). Of course payment must be in proportion to the effort devoted to the research project. Available online at http://grants.nih.gov/grants/policy/fy2012_salary_cap_faqs.htm.

14. S. A. Bunton and W. T. Mallon, "The Continued Evolution of Faculty Appointment and Tenure Policies at U.S. Medical Schools," *Academic Medicine* 82, no. 3 (March 2007), p. 282.

15. NSF does allow payment of full salaries for other personnel such as postdocs and technicians.

16. Bruce Alberts, "Overbuilding Research Capacity," *Science* 329 (September 10, 2010), p. 1257.

17. http://www.whitehouse.gov/omb/circulars_a021_2004/.

18. Indirect cost claims are based on the assignment of space, in other words, unused research space cannot be included. Circular A-21 also has several countervailing provisions, requiring for example that institutions provide at least 25 percent of the equity in cases in which debt financing exceeds $1 million. In addition, NIH staff knowledgeable about OMB Circular A-21 have expressed the view that while the effects of such perverse incentives are possible in theory, they may not be widespread. Personal communication, October 2011.

19. See Stephan, *How Economics Shapes Science*, pp. 161–63. It should be noted, however, that beginning in 2011, lawsuits began to be filed by former law students against an increasing number of U.S. law schools, alleging fraud in the career data provided by these schools. See Katherine Mangan, "Law Schools on the Defensive Over Job-Placement Data," *Chronicle of Higher Education*, October 16, 2011. Available online at http://chronicle.com/article/Crisis-of-Confidence-in-Law/129425/?key=SzgmclZmNCsRNH00YG0SZDpXPXRqZk4kZ3NJPn5wblpRFw%3D%3D.

20. After previous efforts failed, postdocs at the large University of California System voted in 2008 to form a systemwide postdoc union affiliated with the United Auto Workers union. For a discussion of differing perspectives on such developments, see Virginia Gewin, "The Spread of Postdoc Unions," *Nature* 467 (October 6, 2010), pp. 739–41.

21. For more detailed discussion of these possible adjustments, see Teitelbaum, "Structural Disequilibria," pp. 644–45.

22. Bruce Alberts, "Overbuilding Research Capacity," *Science* 329 (September 10, 2010), p. 1257.

23. Biomedical Research Workforce Working Group Report, Advisory Committee to the Director, National Institutes of Health, June 14, 2012, pp. 11, 39–40. Available online at http://acd.od.nih.gov/Biomedical_research_wgreport.pdf.

24. Biomedical Research Workforce Working Group Report, 2012, pp. 11, 42–44.

25. Biomedical Research Workforce Working Group Report, 2012, pp. 9, 36.

26. Biomedical Research Workforce Working Group Report, pp. 10–11, 38–39.

27. Full disclosure: the author has been responsible for an Alfred P. Sloan Foundation grants program that supported the creation of these degrees.

28. National Research Council, Board on Higher Education and Workforce, Policy and Global Affairs, Committee on Enhancing the Master's Degree in the Natural Sciences, *Science Professionals: Master's Education for a Competitive World* (Washington, DC: National Academies Press, 2008). Available online at http://www.nap.edu/catalog.php?record _id=12064. A second committee of the National Research Council also strongly endorsed the PSM initiative in 2012. See National Research Council, Board on Higher Education and Workforce, Policy and Global Affairs, Committee on Research Universities, *Research Universities and the Future of America* (Washington, DC: National Academies Press, 2012), pp. 137–57. Available online at http://www.nap.edu/catalog.php?record_id=13299.

29. Business Higher Education Forum, "Aligning Higher Education STEM Production with Workforce Demand through Professional Master's Degrees," BHEF Issue Brief, June 2011 (Washington, DC: Business Higher Education Forum, 2011). Available online at http:// www.strategicedsolutions.org/sites/default/files/pdfs/Issue_Brief_PSM_Programs.pdf. Council on Competitiveness, *Innovate America* (Washington, DC: Council on Competitiveness, 2005), p. 51. Available online at http://www.compete.org/images/uploads/File/ PDF%20Files/NII_Innovate_America.pdf.

30. Council of Graduate Schools, *Professional Science Master's: A Council of Graduate Schools Guide to Establishing Programs* (Washington, DC: Council of Graduate Schools, 2011). Available online at https://netforum.avectra.com/eweb/shopping/shopping.aspx?pager =1&site=cgs&prd_key=a1b35a0d-faf0-406d-a4e8-4eb27423109a.

31. National Governors Association, "The Professional Science Master's Degree: Meeting the Skills Needs of Innovative Industries." Available online at http://www.nga.org/cms/home/ nga-center-for-best-practices/center-publications/page-ehsw-publications/col2-content/ main-content-list/the-professional-science-masters.html. National Conference of State Legislatures, "Growing the 21st Century Workforce: The Professional Science Master's Degree," (Denver: National Conference of State Legislatures, 2012). Available online at http://www.ncsl.org/issues-research/educ/professional-science-masters-psm.aspx.

32. The NSF provided such support for two years, FY2009 and FY2010. See http://www .nsf.gov/publications/pub_summ.jsp?WT.z_pims_id=503428&ods_key=nsf09607. For comments from the NIH Director's Biomedical Research Workforce Working Group, see Biomedical Research Workforce Working Group Report, 2012, pp. 3, 34.

33. Up-to-date information on this degree is available at www.sciencemasters.com.

34. Renamed in February 2011. Previously known as the Division of Science Resources Statistics (SRS).

35. Biomedical Research Workforce Working Group Report, 2012, pp. 12, 44.
36. Full disclosure: the author was responsible for a series of grants that helped to create and support this project while at the Alfred P. Sloan Foundation.
37. Demetrios G. Papedemetriou, Doris Meissner, Marc R. Rosenblum, and Madeleine Sumption, *Harnessing the Advantages of Immigration for a 21st Century Economy: A Standing Commission on Labor Markets, Economic Competitiveness, and Immigration* (Washington, DC: Migration Policy Institute, 2009). Available online at http://www.migrationpolicy.org/pubs/StandingCommission_May09.pdf.
38. See Additional Statement on this topic by the author in "Breaking the Immigration Stalemate: From Deep Disagreements to Constructive Proposals," A Report from the Brookings-Duke Immigration Policy Roundtable. Conveners: William Galston, Noah Pickus, and Peter Skerry (Brookings Institution and Kenan Institute of Ethics, 2009), p. 28. Available online at http://www.brookings.edu/ and http://kenan.ethics.duke.edu/.
39. National Research Council, Office of Scientific and Engineering Personnel, *Forecasting Demand and Supply of Doctoral Scientists and Engineers: Report of a Workshop on Methodology* (Washington, DC: National Academies Press, 2000), p. 1. The 2000 workshop was chaired by Daniel McFadden, who that same year was awarded the Nobel Memorial Prize in Economics.
40. Thousands of would-be asylum claimants (mostly from Iraq, Afghanistan, Iran, Eastern Europe, and Africa) made their way across France to Sangatte, the French town near Calais at the French terminus of the Channel Tunnel, with the goal of smuggling themselves onto trains to the UK and then filing asylum claims. In response to humanitarian concerns about the numbers of such would-be migrants and political criticism from residents and leaders in Sangatte, the French Red Cross opened a refugee center in Sangatte to provide accommodation for the migrants. The existence of this facility, and the inability or unwillingness of the French government to control it, then stimulated mutual recriminations between the UK and French governments and a firestorm in the British press. The controversy culminated in a directive to the French government from the European Commission that it must limit access by asylum-seekers to cross-Channel trains and close the French Red Cross reception center in Sangatte. For an explanation and justification of the actions taken by the French Red Cross in Sangatte, see article by the editor-in-chief of *Croix-Rouge*, the magazine of the French Red Cross: Pierre Kremer, "Sangatte: A Place of Hope and Despair," *RCRC: The Magazine of the International Red Cross and Red Crescent*, 2002. Available online at http://www.redcross.int/EN/mag/magazine2002_2/sangatte.html.
41. Ray Marshall, *Value-Added Immigration: Lessons for the United States from Canada, Australia, and the United Kingdom* (Washington, DC: Economic Policy Institute, 2011), p. 125.
42. United Kingdom, Migration Advisory Committee, "Skilled, Shortage, Sensible: Review of Methodology" (London: Migration Advisory Committee, March 2010), p. 18. Available online at http://www.ukba.homeoffice.gov.uk/sitecontent/documents/aboutus/workingwithus/mac/review-of-methodology/0310/mac-review--of-methodology?view=Binary.

Appendix A: Controversy about the Meaning of Sputnik

1. Heppenheimer, "How America Chose Not to Beat Sputnik into Space pp. 44–48.
2. Heppenheimer," How America Chose Not to Beat Sputnik into Space," p. 44.

Appendix B: Evolution of the National Institutes of Health

1. This function remained at NIH until 1972, when it was moved to the Food and Drug Administration.
2. Office of NIH History, National Institutes of Health, "A Short History of the National Institutes of Health." Available online at http://history.nih.gov/exhibits/history/index .html.

Appendix C: "A Nation at Risk" and the Sandia Critique

1. National Commission on Excellence in Education, *A Nation at Risk: The Imperative for Educational Reform*, A Report to the Nation and the Secretary of Education, U.S. Department of Education, April 1983. Available online at http://www2.ed.gov/pubs/NatAtRisk/ risk.html.
2. Terrel H. Bell, *Thirteenth Man: A Reagan Cabinet Memoir* (New York: Free Press, 1988), cited in Gerald W. Bracey, "April Foolishness: The 20th Anniversary of 'A Nation at Risk," *The Phi Delta Kappan* 84, no. 8 (April 2003), pp. 616–21, p. 616. Available online at http://www .jstor.org/stable/204404372003.
3. Bracey, "April Foolishness," pp. 616–21. Available online at http://www.jstor.org/stable/ 20440437.
4. See, for example: Bracey, "April Foolishness"; L. Stedman and M. Smith, "Recent Reform Proposals for American Education," *Contemporary Education Review* 2, no. 2 (2003), 85–104; L. Stedman, "The Sandia Report and U.S. Achievement: An Assessment," *Journal of Educational Research* 87, no. 3 (January–February 1994), 133–46.
5. A frequently cited example can be found in Julie A. Miller, "Report Questioning 'Crisis' in Education Triggers an Uproar," *Education Week*, October 9, 1991. Undersecretary of Education David Kearns (formerly the CEO of Xerox Corporation) denounced the Sandia report and allegedly argued that it should be "buried," an allegation Kearns denied. The draft Sandia report was forwarded for review to other federal agencies, including the National Science Foundation and the Department of Education. Without drawing any conclusions, it is interesting to note that the peer reviewer at the National Science Foundation was Peter A. House, who by 1990 had already been playing a prominent role in sounding alarms about looming "shortfalls of scientists and engineers."
6. Stedman, "The Sandia Report and U.S. achievement," citation is on p. 144.
7. Bracey "April Foolishness," p. 617.

INDEX

Abelson, Philip, 45
Abraham, Spencer, 58, 111, 234
Abramoff, Jack, 58, 102–105, 237
acceleration of funding, effects of, 140, 199,
    202, 204–205, 211. *See also* deceleration
    of funding, effects of
Accenture, 89, 91
Advisory Committee to the Director, NIH,
    206, 244, 251–252. *See also* National
    Institutes of Health (NIH)
advocacy research, 6
aerospace industry, 40, 81–82, 129
Afghanistan, 73, 82, 198, 253
agribusiness industry, 98
AILA, 57, 238–239. *See also* American
    Business For Legal Immigration (ABLI);
    American Immigration Lawyers Asso-
    ciation (AILA); Center for American
    Progress (CAP); Compete America;
    Eisen, Jenifer; lobbying activities;
    National Association of Manufacturers
    (NAM)
Air Force, 38, 245
Al-Qaeda attacks on U.S., 109–110, 112. *See
    also* September 2001 attacks on New
    York and Washington
Al-Shehhi, Marwan, 112. *See also* Immigra-
    tion and Naturalization Service (INS);
    September 2001 attacks on New York
    and Washington
alarm/boom/bust cycles, 4–6, 55, 69–72, 174,
    194
Alberts, Bruce, 201, 251–252
Alberty, Robert A., 49
Alliance for Science and Technology
    Research in America (ASTRA) , 62,
    248. *See also* ASTRA
America COMPETES Act, 19, 67–68,
    131–132
America's New Deficit, 93, 234
American Association for the Advancement
    of Science (AAAS), 59–60, 116, 231, 243
American Association of Engineering
    Societies (AAES), 116

American Bar Association (ABA), 114, 116,
    238
American Business For Legal Immigration
    (ABLI), 57–58, 95, 237. *See also* Ameri-
    can Immigration Lawyers Association
    (AILA); Compete America; Eisen,
    Jenifer; National Association of
    Manufacturers (NAM)
American Civil Liberties Union (ACLU),
    105–107. *See also* Left-Right lobbying
    coalitions
American Community Survey (ACS), 145,
    152
American Competitiveness in the Twenty-
    first Century Act of 2000, 58
American Immigration Lawyers Association
    (AILA), 57–58, 105, 238–239. *See also*
    AILA; American Business For Legal
    Immigration (ABLI); Center for
    American Progress (CAP); Eisen,
    Jenifer; Fitz, Marshall; Left-Right
    lobbying coalitions; National
    Association of Manufacturers (NAM)
American Institute of Physics (AIP), 30, 42, 71
American Jewish Committee (AJC), 108
American Medical Association (AMA), 114,
    116
American Recovery and Reinvestment Act
    (ARRA), 19, 32–33, 54, 63, 68
Americans for Tax Reform, 103, 105–106,
    108. *See also* Norquist, Grover
Anderson, Stuart, 92, 94, 111. *See also*
    Abraham, Spencer; Cato Institute;
    CIPRIS, Coordinated Interagency
    Partnership Regulating International
    Students; Help Wanted; ITAA, Infor-
    mation Technology Association of
    America; NAFSA:Association of
    International Educators
Apollo program, 36, 41, 43, 45–46, 48, 54, 60,
    142, 229
applied research, 47, 143, 176, 190, 192
Arab oil boycott, 77. *See also* oil crises; oil
    prices

Johnson, Clarence "Kelly," 37. *See also* Lockheed Corporation; Skunk Works; U-2
Johnson, Lyndon, 40–41, 43, 45
joint funding, 161–163, 165, 170
joint production, 161, 163, 170
Jordan, Barbara L., 108. *See also* Commission on Immigration Reform, U.S.
junior investigators, 199, 213. *See also* young scientists

K-12 education, 16, 99, 101, 116
Kagan, Elena, 96, 235. *See also* Fernandes, Julie A.
Kaiser, David, 29, 32, 227–228, 230
keiretsu, 73
Kennedy, Edward M., 67, 96, 99
Kennedy, John F., 43, 73, 229
Kennedy-McCain immigration bill, 96, 99. *See also* comprehensive immigration reform (CIR)
Killewald, Alexandra, 22–23, 126, 139–140, 169, 227, 243–245, 248, 251
Killian, James, 35–37
Kirschstein, Ruth L., 162
Korol, Alexander, 34, 228

labor certification (LC), 91
labor condition application (LCA), 91
labor shortage and surplus, conflicting definitions of, 118–120, 129–132, 138–139, 240; cyclical, 122, 133; dynamic, 121–122; static, 121; structural, 121–122, 133. *See also* shortages of scientists and engineers; shortfalls of scientists and engineers; surpluses of scientists and engineers
Laika, 41. *See also* Sputnik
Land, Edwin, 37
Lederman, Leon, 59, 231. *See also* Science: The End of the Frontier
left wing politics, 88, 105–109, 117, 184, 234, 239. *See also* Left-Right lobbying coalitions; right wing politics
Left-Right lobbying coalitions, 88, 105–109, 117, 184, 234, 239. *See also* Eisen, Phyllis; Gilder, Richard; National Immigration Forum; Swartz, Rick
Lenovo, 95
Levchin, Max, 98
leverage, financial in universities, 200–201, 206

liberal arts, 182, 184
libertarian, right- and left-, 58, 92, 106–108, 234
limited information about STEM careers, 170, 202, 239
LinkedIn, 98
lobbying activities, 5, 55, 57–58, 60, 62, 67, 69, 88, 94–99, 101–106, 108–109, 112–113, 116, 204–205, 232, 235–239
Lobbying Disclosure Act, 97
lobbyists, 95–96, 102–103, 106, 237–238
Lockheed Corporation, 37
Lockheed Martin Corporation, 8, 13, 15
looming shortfalls, 4, 60, 209. *See also* Atkinson, Richard; Bloch, Erich; House, Peter W.; Policy Research and Analysis (PRA), NSF
lottery characteristics of science awards, 140
Lowell, B. Lindsay, 21, 227, 243–244
Lysenko, 35

Mack, Connie, 61
Majority Group, 102. *See also* lobbying activities; lobbyists
Manhattan Institute, 107
Manhattan Project, 29, 159, 250
Manning, Bradley or Chelsea, 123
Mansfield Amendment, 36, 48–49, 230
Mansfield, Mike, 48
manufacturing, 8, 29, 73–74, 129, 173, 177, 241, 248
Marine Hospital Service (MHS), 219
Martin, Philip, 129, 242
Martinez, Mel, 96–97, 99. *See also* Hagel-Martinez immigration bill
Master's degrees, 139, 143–144, 156, 252. *See also* Professional Science Master's degree (PSM)
Mayer, Marissa, 98
MBA, 137, 245
McCain, John, 96, 99. *See also* Kennedy-McCain immigration bill
McKinsey Global Institute, 135
Medaris, John B., 218. *See also* Explorer 1 program; von Braun, Wernher
Medicare, 122, 196
Meissner, Doris, 111, 253. *See also* Berez, Maurice; Immigration and Naturalization Service (INS)
Meredith Advocacy Group, 98. *See also* lobbying activities; lobbyists